Contents

Part 1: The Respiratory Pathway, Lungs, Thoracic Wall and Diaphragm

Part 2: The Heart and Great Veins of the Neck

Part 3: The Vertebral Canal and its Contents

Anatomy for Anaesthetists

Anatomy for Anaesthetists

HAROLD ELLIS

CBE, MA, DM, MCh, FRCS, FRCP, FRCOG, FACS (Hon)
Clinical Anatomist, Guy's, King's and St Thomas' School of
Biomedical Sciences;
Emeritus Professor of Surgery, Charing Cross
and Westminster Medical School, London;

W   well

This edition first published 2014 © 1963, 1969, 1977, 1983, 1988, 1993, 1997, 2004 by Blackwell Science Ltd. 2014 © John Wiley & Sons, Ltd

Registered office: John Wiley & Sons, Ltd, The Atrium, Southern Gate, Chichester, West Sussex, PO19 8SQ, UK

Editorial offices: 9600 Garsington Road, Oxford, OX4 2DQ, UK
111 River Street, Hoboken, NJ 07030-5774, USA

For details of our global editorial offices, for customer services and for information about how to apply for permission to reuse the copyright material in this book please see our website at www.wiley.com/wiley-blackwell

Library of Congress Cataloging-in-Publication Data

Ellis, Harold, 1926– author.
  Anatomy for anaesthetists / Harold Ellis, Andrew Lawson. – Ninth edition.
     p. ; cm.
  Includes bibliographical references and index.
  ISBN 978-1-118-37598-3 (cloth : alk. paper) – ISBN 978-1-118-37594-5 –
ISBN 978-1-118-37595-2 (emobi) – ISBN 978-1-118-37596-9 (epdf) –
ISBN 978-1-118-37597-6 (epub)
  I. Lawson, Andrew (Medical ethicist), author.   II. Title.
  [DNLM: 1. Anatomy.   2. Anesthesia. QS 4]
  QM23.2
  611′.0024617–dc23
                                          2013024798

A catalogue record for this book is available from the British Library.

Wiley also publishes its books in a variety of electronic formats. Some content that appears in print may not be available in electronic books.

Cover image: Science Photo Library image: © POGEE/SCIENCE PHOTO LIBRARY
Color-enhanced x-ray showing endotracheal intubation
Cover design by Meaden Creative

Set in 9/12pt Palatino by Aptara® Inc., New Delhi, India
Printed and bound in Singapore by Markono Print Media Pte Ltd

1   2013

Part 4: The Peripheral Nerves

Part 5: The Autonomic Nervous System

Part 6: The Cranial Nerves

Part 7: Miscellaneous Zones of Interest

Part 8: The Anatomy of Pain

Preface to the Ninth (Jubilee) Edition

I little thought, when I wrote the Introduction to the first edition of *Anatomy for Anaesthetists* for its publication in 1963, that one day I would be composing the introduction to its ninth (Jubilee) edition in 2013! The place of anatomy in anaesthetic practice is often considered merely as a prerequisite for the safe practice of local anaesthetic blocks. However, it is also important in understanding the anatomy of the airways, the functions of the lung and circulation, monitoring of neuromuscular blocks, long-term pain control and many other aspects of practical anaesthesia. This book is not intended to be a textbook of regional anaesthetic techniques; there are many excellent texts that cover this important field. However, it is a textbook written for anaesthetists, keeping in mind the special requirements of their daily practice. The anaesthetist requires a particularly specialized knowledge of anatomy. Some regions of the body, for example the respiratory passages, the major veins and the peripheral nerves, the anaesthetist must know with an intimacy of detail that rivals or even exceeds that of the surgeon; other areas can be all but ignored.

The first edition was written in collaboration with that talented medical artist, the late Margaret McLarty. I was then joined by my anaesthetic colleague at Westminster Medical School, Professor Stanley Feldman, as co-author. Dr William Harrop-Griffiths contributed much help on peripheral nerve blocks to the eighth edition. I am now delighted to have Dr Andrew Lawson as my collaborator. He has already added the important chapter on the anatomy of pain to the last two editions. He is not only an anaesthetist but also an expert in pain management and brings his expertise in this field in describing important clinical applications of anatomy to this aspect of anaesthetic practice.

In this ninth edition, we have carefully revised and expanded the text and have added new illustrations, including modern imaging techniques that are of particular interest to the anaesthetist. We hope this book will continue to serve anaesthetists as it has done over the past 50 years.

Harold Ellis
July 2013

Foreword to the First Edition

The anaesthetist faced with higher examinations is confronted with the problem of how far he should delve into the many related specialities; and owing to constant change of emphasis, the answers will never be final. Professor Ellis, as a surgeon who has interested himself in our speciality for a number of years, here sets out boldly the anatomy he believes the present-day young anaesthetist should be familiar with when confronting the examiners. The choice of material, and its presentation, are full of commonsense, so that the book is attractive to the established anaesthetist in his daily work, and for occasional reading.

I have had the privilege of working with both authors for a number of years, and I can think of no pair better fitted to highlight the essentials of anatomy for the anaesthetist. Professor Ellis is an outstanding teacher, and in the field of medical artists Miss McLarty is distinguished by her flair for stressing points of anatomical importance to the clinician. I persuaded them to collaborate on a series of articles which recently appeared in *Anaesthesia*. Though these have been added to considerably they form the basis of this book.

Professor Sir Robert Macintosh

Introduction to the First Edition

The anaesthetist requires a peculiarly specialized knowledge of anatomy. Some regions of the body – the nerve pathways and respiratory passages for example – he must know with an intimate detail which rivals that of the surgeon; other areas he may all but ignore. As far as we know this book is the first to be designed with his particular needs in mind. It should prove of value to examination candidates and we hope also to the practical anaesthetist in his day to day work.

Harold Ellis, London
Margaret McLarty, Oxford
1963

Acknowledgements to the Ninth (Jubilee) Edition

The first two editions of this textbook were prepared in collaboration with that skilled medical artist Miss Margaret McLarty. The illustrations for the sixth edition were almost all drawn or redrafted by Rachel Chesterton; we thank her for the excellent way in which they have been executed. Further illustrations for the seventh, eighth and this edition were prepared by Jane Fallows with great skill. Some of the figures have been reproduced from *Clinical Anatomy*, 13th edition. Thanks to Dr Charles Gaucci, Dr Ron Cooper, Dr Nick Morgan Hughes and Dr Vlademir Gorelov for providing images for use in this book.

Part 1

The Respiratory Pathway, Lungs, Thoracic Wall and Diaphragm

Anatomy for Anaesthetists, Ninth Edition. Harold Ellis and Andrew Lawson.
© 2014 John Wiley & Sons, Ltd. Published 2014 by John Wiley & Sons, Ltd.

The mouth

The mouth is made up of the vestibule and the mouth cavity, the former communicating with the latter through the aperture of the mouth.

The *vestibule* is formed by the lips and cheeks without, and by the gums and teeth within. An important feature is the opening of the *parotid duct* on a small papilla opposite the 2nd upper molar tooth. Normally the walls of the vestibule are kept together by the tone of the facial muscles; a characteristic feature of a facial (VII) nerve paralysis is that the cheek falls away from the teeth and gums, enabling food and drink to collect in, and dribble out of, the now patulous vestibule.

The *mouth cavity* (Fig. 1) is bounded by the alveolar arch of the maxilla and the mandible, and teeth in front, the hard and soft palate above, the anterior two-thirds of the tongue and the reflection of its mucosa forwards onto the mandible below, and the oropharyngeal isthmus behind.

The mucosa of the floor of the mouth between the tongue and mandible bears the median *frenulum linguae*, on either side of which are the orifices of the submandibular salivary glands (Fig. 2). Backwards and outwards from these ducts extend the sublingual folds that cover the sublingual glands on each side (Fig. 3); the majority of the ducts of these glands open as a series of tiny orifices along the overlying fold, but some drain into the duct of the submandibular gland (Wharton's duct).

Inspect your mouth in a mirror. Elevate your tongue, then press on one or the other side onto your submandibular gland beneath the angle of the jaw. You will see a jet of saliva emerge from the orifice of the submandibular duct at the tip of the sublingual fold. While about it, pull your cheek laterally with a finger, press on your parotid gland on that side and observe a jet of saliva emerge from the parotid duct, which lies at the level of your 2nd upper molar tooth.

The palate

The *hard palate* is made up of the palatine processes of the maxillae and the horizontal plates of the palatine bones. The mucous membrane covering the hard palate is peculiar in that the stratified squamous mucosa is closely connected to the underlying periosteum, so that the two dissect away at operation as a single sheet termed the mucoperiosteum. This is thin in the midline, but thicker more laterally owing to the presence of numerous small palatine salivary glands, an uncommon but well-recognized site for the development of mixed salivary tumours.

The *soft palate* hangs like a curtain suspended from the posterior edge of the hard palate. Its free border bears the *uvula* centrally and blends on either side with the pharyngeal wall. The anterior aspect of this curtain faces the mouth cavity and is covered by a stratified squamous epithelium. The posterior aspect is part of the nasopharynx and is lined by a ciliated columnar epithelium under which is a thick stratum of mucous and serous glands embedded in lymphoid tissue.

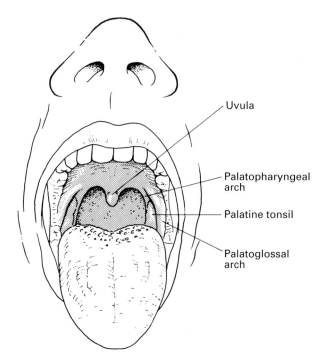

Uvula

Palatopharyngeal arch

Palatine tonsil

Palatoglossal arch

Fig. 1 View of the open mouth with the tongue depressed.

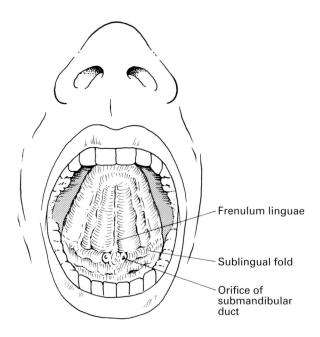

Frenulum linguae

Sublingual fold

Orifice of submandibular duct

Fig. 2 View of the open mouth with the tongue elevated.

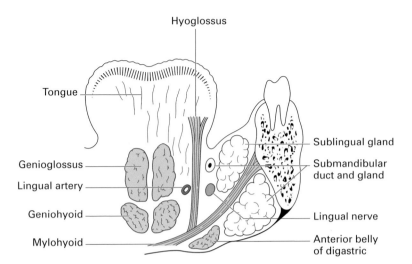

Fig. 3 Coronal section through the floor of the mouth.

The 'skeleton' of the soft palate is a tough fibrous sheet termed the *palatine aponeurosis*, which is attached to the posterior edge of the hard palate. The aponeurosis is continuous on each side with the tendon of tensor palati and may, in fact, represent an expansion of this tendon.

The *muscles of the soft palate* are five in number: the tensor palati, the levator palati, the palatoglossus, the palatopharyngeus and the musculus uvulae (see Fig. 13).

The *tensor palati* arises from the scaphoid fossa at the root of the medial pterygoid plate, from the lateral side of the Eustachian cartilage and the medial side of the spine of the sphenoid. Its fibres descend laterally to the superior constrictor and the medial pterygoid plate to end in a tendon that pierces the pharynx then loops medially around the hook of the hamulus to be inserted into the palatine aponeurosis. Its action is to tighten and flatten the soft palate.

The *levator palati* arises from the undersurface of the petrous temporal bone and from the medial side of the Eustachian tube, enters the upper surface of the soft palate and meets its fellow of the opposite side. It elevates the soft palate.

The *palatoglossus* arises in the soft palate, descends in the palatoglossal fold and blends with the side of the tongue. It approximates the palatoglossal folds.

The *palatopharyngeus* descends from the soft palate in the palatopharyngeal fold to merge into the side wall of the pharynx: some fibres become inserted along the posterior border of the thyroid cartilage. It approximates the palatopharyngeal folds.

The *musculus uvulae* takes origin from the palatine aponeurosis at the posterior nasal spine of the palatine bone and is inserted into the uvula. Injury to the cranial root of the accessory nerve, which supplies this muscle via the vagus nerve, results in the uvula becoming drawn across and upwards towards the opposite side.

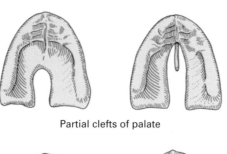

Partial clefts of palate

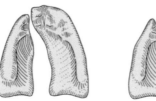

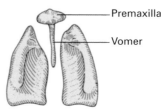

Premaxilla

Vomer

Unilateral complete cleft palate   Bilateral complete cleft palate

Fig. 4 Types of cleft palate deformity.

The tensor palati is innervated by the mandibular branch of the trigeminal nerve via the otic ganglion (see page 274). The other palatine muscles are supplied by the pharyngeal plexus of the vagus nerve, which transmits cranial fibres of the accessory nerve via the vagus.

The palatine muscles help to close off the nasopharynx from the mouth in deglutition and phonation. In this, they are aided by contraction of the upper part of the superior constrictor, which produces a transverse ridge on the back and side walls of the pharynx at the level of the 2nd cervical vertebra termed the *ridge of Passavant*.

Paralysis of the palatine muscles results (just as surely as a severe degree of cleft palate deformity) in a typical nasal speech and in regurgitation of food through the nose.

Cleft palate

The palate develops from a central premaxilla and a pair of lateral maxillary processes: the former usually bears all four (occasionally only two) of the incisor teeth. All degrees of failure of fusion of these three processes may take place. There may be a complete cleft, which passes to one or both sides of the premaxilla; in the latter case, the premaxilla prolapses forwards to produce a marked deformity. Partial clefts of the posterior palate may involve the uvula only (bifid uvula), involve the soft palate or encroach into the posterior part of the hard palate (Fig. 4).

The nose

The nose is divided anatomically into the external nose and the nasal cavity.

The *external nose* is formed by an upper framework of bone (made up of the nasal bones, the nasal part of the frontal bones and the frontal processes of the maxillae), a series of cartilages in the lower part, and a small zone of fibrofatty tissue that forms the lateral margin of the nostril (the ala). The cartilage of the nasal septum constitutes the central support of this framework.

The *cavity of the nose* is subdivided by the nasal septum into two quite separate compartments that open to the exterior by the *nares* and into the nasopharynx by the posterior nasal apertures or *choanae*. Immediately within the nares is a small dilatation, the *vestibule*, which is lined in its lower part by stiff straight hairs.

Each side of the nose presents a roof, a floor and a medial and lateral wall.

The *roof* first slopes upwards and backwards to form the bridge of the nose (the nasal and frontal bones), then has a horizontal part (the cribriform plate of the ethmoid), and finally a downward-sloping segment (the body of the sphenoid).

The *floor* is concave from side to side and slightly so from before backwards. It is formed by the palatine process of the maxilla and the horizontal plate of the palatine bone.

The *medial wall* (Fig. 5) is the nasal septum, formed by the septal cartilage, the perpendicular plate of the ethmoid and the vomer. Deviations of the septum are very common; in fact, they are present to some degree in about 75% of the adult population. Probably nearly all are traumatic in

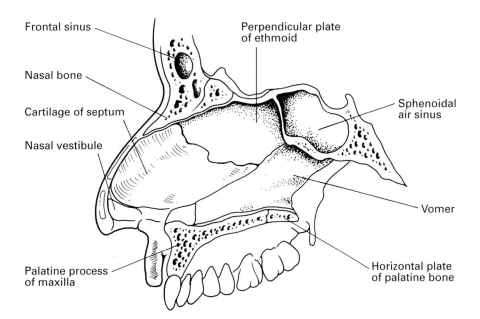

Frontal sinus

Nasal bone

Cartilage of septum

Nasal vestibule

Perpendicular plate of ethmoid

Sphenoidal air sinus

Vomer

Palatine process of maxilla

Horizontal plate of palatine bone

Fig. 5 The septum of the nose.

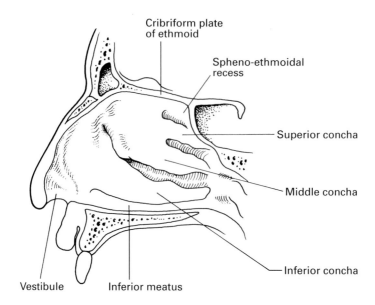

Cribriform plate
of ethmoid

Spheno-ethmoidal
recess

Superior concha

Middle concha

Inferior concha

Vestibule   Inferior meatus

Fig. 6 The lateral wall
of the right nasal cavity.

origin, and result from quite minor injuries in childhood or even at birth. The deformity does not usually manifest itself until the second dentition appears, when rapid growth in the region produces deflections from what had been an unrecognized minor dislocation of the septal cartilage. Males are more commonly affected than females, a distribution which would favour this traumatic theory. Both nostrils may become blocked, either from a sigmoid deformity of the cartilage or from compensatory hypertrophy of the conchae on the opposite side. The deviation is nearly always confined to the anterior part of the septum.

The *lateral wall* (Fig. 6) has a bony framework made up principally of the nasal aspect of the ethmoidal labyrinth above, the nasal surface of the maxilla below and in front, and the perpendicular plate of the palatine bone behind. This is supplemented by the three scroll-like conchae (or turbinate bones), each arching over a *meatus*. The upper and middle *conchae* are derived from the medial aspect of the ethmoid labyrinth; the inferior concha is a separate bone.

Onto the lateral wall open the orifices of the paranasal sinuses (see page 9) and the nasolacrimal duct; the arrangement of these orifices is shown in Fig. 7.

The sphenoid sinus opens into the *spheno-ethmoidal recess*, a depression between the short superior concha and the anterior surface of the body of the sphenoid. The posterior ethmoidal cells drain into the superior meatus. The middle ethmoidal cells bulge into the middle meatus to form an elevation, termed the *bulla ethmoidalis*, onto which they open. Below the bulla is a cleft, the *hiatus semilunaris*, into which opens the ostium of the maxillary sinus. The hiatus semilunaris curves forwards in front of the bulla ethmoidalis as a passage termed the *infundibulum*, which drains the anterior ethmoidal air cells. In about 50% of cases the frontal sinus drains into the

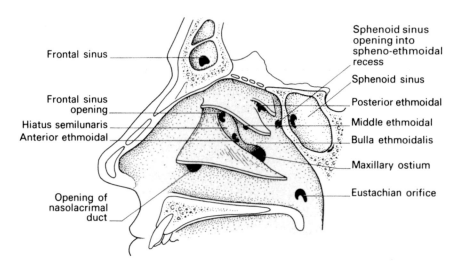

Fig. 7 The lateral wall of the right nasal cavity; the conchae have been partially removed.

infundibulum via the frontonasal duct. In the remainder, this duct opens into the anterior extremity of the middle meatus.

The nasolacrimal duct drains tears into the anterior end of the inferior meatus in solitary splendour.

The paranasal sinuses

The paranasal air sinuses comprise the maxillary, sphenoid, frontal and ethmoidal sinuses. They are, in effect, the outpouchings from the lateral wall of the nasal cavity into which they drain; they all differ considerably from subject to subject in their size and extent, and they are rarely symmetrical. There are traces of the maxillary and sphenoid sinuses in the newborn; the rest become evident about the age of 7 or 8 years in association with the eruption of the second dentition and lengthening of the face. They only become fully developed at adolescence.

The *maxillary sinus* (the antrum of Highmore) is the largest of the sinuses. It is pyramid-shaped, and occupies the body of the maxilla (Fig. 8). The base of this pyramid is the lateral wall of the nasal cavity and its apex points laterally towards the zygomatic process.

The floor of the sinus extends into the alveolar process of the maxilla, which lies approximately 1.25 cm below the level of the floor of the nose. Bulges in the floor are produced by the roots of at least the 1st and 2nd molars; the number of such projections is variable and may include all the teeth derived from the maxillary process, i.e. the canine, premolars and molars. The floor may actually be perforated by one or more of the roots.

The roof is formed by the orbital plate of the maxilla, which bears the canal of the infra-orbital branch of the maxillary nerve. Medially, the antrum drains into the middle meatus; the ostium is situated high up on

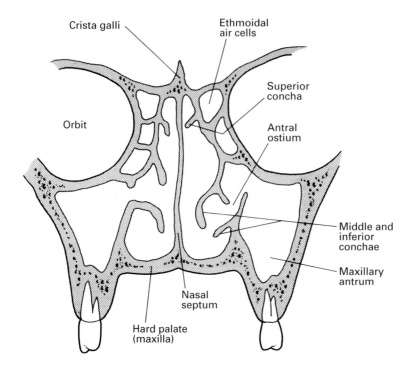

Crista galli

Ethmoidal
air cells

Superior
concha

Orbit

Antral
ostium

Middle and
inferior
conchae

Maxillary
antrum

Nasal
septum

Hard palate
(maxilla)

Fig. 8 The maxillary
sinus in coronal
section.

this wall and is thus inefficiently placed from the mechanical point of view. Drainage from this sinus is therefore dependent on the effectiveness of the cilia lining its wall. There may be one or more accessory openings from the antrum into the middle meatus.

The *sphenoid sinuses* lie side by side in the body of the sphenoid. Occasionally, they extend into the basisphenoid and the clinoid processes. They are seldom equal in size, and the septum between them is usually incomplete. They open into the spheno-ethmoidal recess.

The *frontal sinuses* occupy the frontal bone above the orbits and the root of the nose. They are usually unequal, and their dividing septum may be incomplete. It is interesting that their extent is in no way related to the size of the superciliary ridges. They drain through the frontonasal duct into the middle meatus.

The *ethmoidal sinuses* or air cells are made up of some 8–10 loculi suspended from the outer extremity of the cribriform plate of the ethmoid and bounded laterally by its orbital plate. They thus occupy the upper lateral wall of the nasal cavity. The cells are divided into anterior, middle and posterior groups by bony septa; their openings have already been described above.

Blood supply

The upper part of the nasal cavity receives its arterial supply from the anterior and posterior ethmoidal branches of the ophthalmic artery, a branch of the internal carotid artery. The sphenopalatine branch of the maxillary artery is distributed to the lower part of the cavity and links up with the septal branch of the superior labial branch of the facial artery on the antero-inferior part of the septum. It is from this zone, just within the vestibule of the nose, that epistaxis occurs in some 90% of cases (Little's area).

A rich submucous venous plexus drains into the sphenopalatine, facial and ophthalmic veins, and through the last links up with the cavernous sinus. Small tributaries also pass through the cribriform plate to veins on the undersurface of the orbital lobe of the brain. These connections account for the potential danger of boils and other infections within and adjacent to the nose.

Nerve supply

The olfactory nerve (I) supplies the specialized olfactory zone of the nose, which occupies an area of some 2 cm^2 in the uppermost parts of the septum and lateral walls of the nasal cavity (see page 247).

The nerves of common sensation are derived from the nasociliary branch of the 1st division of the trigeminal nerve (V′) and also from the 2nd, or maxillary, division (V″). These nerves are considered fully in Chapter 6, but may conveniently be summarized here.

1 The *septum* (Fig. 9) is supplied, in the main, by the nasopalatine nerve, derived from V″ via the pterygopalatine ganglion. The posterosuperior corner receives branches of the medial posterosuperior nasal nerves from the same source, and the anterior part of the septum is supplied

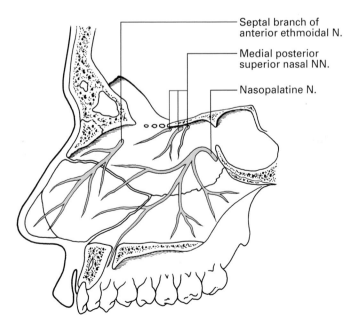

Septal branch of anterior ethmoidal N.

Medial posterior superior nasal NN.

Nasopalatine N.

Fig. 9 The nerve supply of the nasal septum.

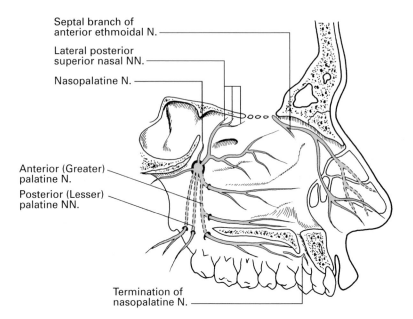

Septal branch of
anterior ethmoidal N.

Lateral posterior
superior nasal NN.

Nasopalatine N.

Anterior (Greater)
palatine N.

Posterior (Lesser)
palatine NN.

Termination of
nasopalatine N.

Fig. 10 The nerve supply of the lateral wall of the nose.

by the septal branches of the anterior ethmoidal nerve (a branch of the nasociliary branch of V').

2 The *lateral wall* (Fig. 10) is innervated in its upper part, in the region of the superior and middle conchae, by the lateral posterior superior nasal nerve. The inferior concha receives branches from the anterior superior alveolar nerve (arising from the maxillary nerve in the infra-orbital canal) and from the anterior (greater) palatine nerve (derived from the pterygopalatine ganglion). The anterior part of the lateral wall, in front of the conchae, is supplied by the anterior ethmoidal branch of the nasociliary nerve. This branch then leaves the nasal cavity between the nasal bone and the upper nasal cartilage to become the external nasal nerve, which supplies the outer aspect of the nose; the anterior ethmoidal nerve thus innervates the cartilaginous tip of the nose on both its inner and outer aspects.

3 The *floor* is supplied in its anterior part by the anterosuperior alveolar nerve and posteriorly by the anterior (greater) palatine nerve.

4 The *vestibule* receives terminal twigs of the infra-orbital branch of the maxillary nerve, which also supplies the skin immediately lateral to, and beneath, the nose.

5 The *paranasal sinuses* are innervated by V' and V''. The maxillary sinus is supplied entirely by the maxillary nerve; its roof by the infra-orbital nerve, floor by the anterior palatine nerve, medial wall by the medial posterosuperior nasal and the anterior (greater) palatine nerves, and the anterior, posterior and lateral walls by the superior alveolar branches. The other sinuses are supplied by the ophthalmic division of V: the ethmoidal and sphenoidal sinuses by the anterior and posterior ethmoidal

nerves, and the frontal sinus by the supra-orbital and supratrochlear nerves.

Structure

The vestibule is lined by a stratified squamous epithelium bearing stiff straight hairs, sebaceous glands and sweat glands. The remainder of the nasal cavity, apart from the small olfactory area, bears tall columnar ciliated cells interspersed with mucus-secreting goblet cells, and forms a continuous epithelial sheet with the mucosa of the nasal sinuses. Beneath the epithelium is a highly vascular connective tissue containing copious lymphoid aggregates and carrying mucous and serous glands. The mucous membrane is thick and velvety over the greater part of the nasal septum and over the conchae. However, it is thin over the septum immediately within the vestibule (where the blood vessels of Little's area show through the mucosa) and also over the meati and the floor of the nose.

The mucosa of the nose, and its accessory sinuses, is closely adherent to the underlying periosteum or perichondrium; surgically, the two layers strip away together and, as in the hard palate, are termed the mucoperiosteum.

The functions of the nose

The nose acts as a respiratory pathway, through which air becomes warmed, humidified and filtered, as the organ of olfaction and as a resonator in speech.

There is a strong inborn reflex to breathe through the nose. This is natural to the survival of babies during suckling. As a result, nasal obstruction may cause gross discomfort; thus, packing the nose after surgery may cause restlessness upon emergence from an anaesthetic, and choanal atresia may cause cyanosis in the newborn. The natural expiratory resistance of the upper airways is of the order of 1–2 cmH$_2$O and can be increased subconsciously to provide a natural form of continuous positive airway pressure. Intubation of the trachea decreases this natural expiratory resistance.

Air passes through the nose, not directly along the inferior meatus, but in a curve through the upper reaches of the nasal cavity. The vascular cavernous plexuses, arranged longitudinally like so many radiator pipes, increase the temperature of the air to that of the body by the time it reaches the nasopharynx. Water, derived partly from the mucous and serous glands, partly from the goblet cells, but mainly by exudation from the mucous surfaces, produces nearly 100% saturation of the inhaled air. Filtration is effected by the blanket of mucus covering the nasal cavity and its related sinuses. The mucus is swept towards the pharynx like a sticky conveyor belt by the action of the cilia and then swallowed. Reflex sneezing also helps rid the nose of irritants.

The blood supply to the nasal mucosa is under reflex control. General warming of the subject produces reflex hyperaemia whereas general cooling results in vasoconstriction. Hence the well-known observation that one's stuffy nose in a hot room clears on going out into the cold air.

A part of the Horner's syndrome produced in a cervical sympathetic block (see page 162) is blockage of the nasal passage on that side as a result of paralysis of sympathetic vasoconstrictor fibres to the nasal mucosa.

CLINICAL NOTE

Nasal intubation

The major nasal air passage lies beneath the inferior concha, and a nasotracheal tube should be encouraged to use this passage by passing it directly backwards along the floor of the nose. Occasionally, the posterior end of the inferior turbinate may be hypertrophied and may offer resistance to the easy passage of the tube. The delicate mucosa of the nose and the posterior pharyngeal wall may easily be torn, and force must never be used in this manoeuvre. Cases are on record of nasal tubes being passed through the mucosa of the posterior pharyngeal wall into the retropharyngeal space and of serious haemorrhage from injury to the posterior ethmoidal vessels, which are branches of the internal carotid artery via the ophthalmic artery and therefore impossible to control by proximal ligation. It can be seen from Fig. 11 that a nasotracheal tube must curve anteriorly as it passes through the nasopharynx. It may be possible to pass a well-curved tube in a 'blind' manner, but more flexible tubes will need assistance if they are to be passed through the vocal cords. Magill's intubating forceps are commonly used for

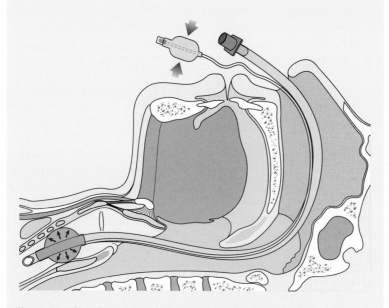

Fig. 11 Nasal intubation; note the curvatures of the tracheal tube.

this purpose. A well-curved and rigid tube may increase the chances of success of attempts at blind nasal intubation, but may also increase the chances of trauma to the anterior tracheal wall. Experienced operators often manipulate the larynx with one hand and intubate with the other. Confirmation of intubation may be visual by laryngoscopy or by capnography (measurement of exhaled CO_2).

The pharynx

The pharynx is a wide muscular tube that forms the common upper pathway of the respiratory and alimentary tracts. Anteriorly, it is in free communication with the nasal cavity, the mouth and the larynx, which conveniently divide it into three parts, termed the nasopharynx, oropharynx and laryngopharynx, respectively (Figs 12, 13). In extent, it reaches from the skull (the basilar part of the occipital bone) to the origin of the oesophagus at the level of the 6th cervical vertebra (C6). Posteriorly, it rests against the cervical vertebrae and the prevertebral fascia.

The nasopharynx

The nasopharynx lies behind the nasal cavity and above the soft palate. It communicates with the oropharynx through the pharyngeal isthmus, which becomes closed off during the act of swallowing (see page 21). On the lateral wall of the nasopharynx, 1 cm behind and just below the

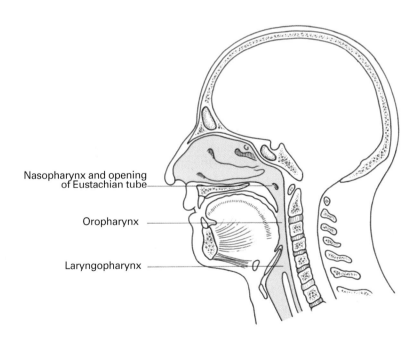

Nasopharynx and opening of Eustachian tube

Oropharynx

Laryngopharynx

Fig. 12 A sagittal section through the head and neck to show the subdivisions of the pharynx.

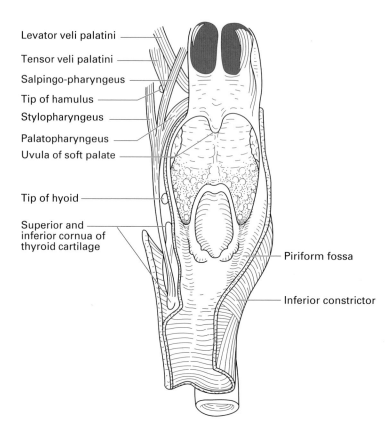

Levator veli palatini

Tensor veli palatini

Salpingo-pharyngeus

Tip of hamulus

Stylopharyngeus

Palatopharyngeus

Uvula of soft palate

Tip of hyoid

Superior and inferior cornua of thyroid cartilage

Piriform fossa

Inferior constrictor

Fig. 13 The pharynx: the posterior wall has been removed and the interior is viewed from behind. On the left side the palatal muscles have been exposed.

inferior nasal concha, lies the pharyngeal opening of the *pharyngotympanic (Eustachian) tube*. The underlying cartilage of the tube produces a bulge immediately behind its opening, termed the *tubal elevation*, and behind this, in turn, is a small depression, the *pharyngeal recess* – fossa of Rosenmüller (Fig. 7).

The *nasopharyngeal tonsil* ('adenoids') lies on the roof and posterior wall of the nasopharynx. It consists of a collection of lymphoid tissue covered by ciliated epithelium and lies directly against the superior constrictor muscle; it has no well-defined fibrous capsule. The lymphoid tissue begins to atrophy at puberty and has all but disappeared by early adult life.

Posterosuperiorly to the nasopharynx lies the sphenoid sinus that separates the pharynx from the sella turcica containing the pituitary gland. This is the basis of the transnasal approach to the pituitary.

The oropharynx

The mouth cavity leads into the oropharynx through the oropharyngeal isthmus, which is bounded by the palatoglossal arches, the soft palate and the dorsum of the tongue (Fig. 1). The oropharynx itself extends in height

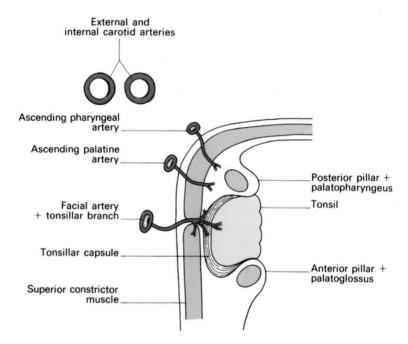

External and
internal carotid arteries

Ascending pharyngeal
artery

Ascending palatine
artery

Facial artery
+ tonsillar branch

Tonsillar capsule

Superior constrictor
muscle

Posterior pillar +
palatopharyngeus

Tonsil

Anterior pillar +
palatoglossus

Fig. 14 Diagram of the tonsil and its surroundings in horizontal section.

from the soft palate to the tip of the epiglottis. Its most important features are the tonsils.

The *palatine tonsils* are the collections of lymphoid tissue that lie on each side in the triangle formed by the palatoglossal and palatopharyngeal arches (the pillars of the fauces), connected across the base by the dorsum of the tongue (Fig. 1). The free surface of each palatine tonsil presents about 12–20 tonsillar pits, and its upper part bears the intratonsillar cleft. This free surface is covered by a stratified squamous epithelium: the unique combination of squamous epithelium with underlying lymphoid tissue renders a section through the tonsil unmistakable under the microscope.

The deep surface of the palatine tonsil may send processes of lymphoid tissue into the dorsum of the tongue, into the soft palate and into the faucal pillars. The palatine tonsil is bounded on this deep aspect by a dense fibrous *capsule* of thickened pharyngeal aponeurosis, which is separated by a film of lax connective tissue from the underlying superior constrictor muscle (Fig. 14). In the absence of inflammation, this capsule enables complete enucleation of the tonsil to be effected. However, after repeated quinsies, the capsule becomes adherent to the underlying muscle and tonsillectomy then requires sharp dissection.

Vascular, lymphatic and nerve supply

The principal blood supply of the palatine tonsil is the tonsillar branch of the facial artery which, accompanied by its two venae comitantes, pierces the superior constrictor muscle to enter the inferior pole of the tonsil. In addition, twigs from the lingual, ascending palatine, ascending pharyngeal and maxillary arteries all add their contributions. Venous drainage passes into the venae comitantes of the tonsillar branch of the facial artery and also into a paratonsillar vein that descends from the soft palate across the outer aspect of the tonsillar capsule to pierce the pharyngeal wall into the pharyngeal venous plexus. It is this vein which is the cause of occasional unpleasant venous bleeding after tonsillectomy. The internal carotid artery, it should be noted, is a precious 2.5 cm away from the tonsillar capsule, and is out of harm's way during tonsillectomy (Fig. 14).

Lymph drainage is to the upper deep cervical nodes, particularly to the jugulo-digastric node (or tonsillar node) at the point where the common facial vein joins the internal jugular vein.

There is a threefold sensory nerve supply:
1 the glossopharyngeal nerve via the pharyngeal plexus;
2 the posterior palatine branch of the maxillary nerve;
3 twigs from the lingual branch of the mandibular nerve.

For this reason, infiltration anaesthesia of the tonsil is more practicable than attempts at nerve blockade.

The palatine and pharyngeal tonsils, together with lymph collections on the posterior part of the tongue and in relation to the Eustachian orifice, form a more or less continuous ring of lymphoid tissue around the pharyngeal entrance, which is termed Waldeyer's ring.

The laryngopharynx

The third part of the pharynx extends from the tip of the epiglottis to the lower border of the cricoid at the level of C6 (Fig. 13). Its anterior aspect faces first the laryngeal inlet, bounded by the aryepiglottic folds, then, below this, the posterior aspects of the arytenoids, and finally the cricoid cartilage. The larynx bulges back into the centre of the laryngopharynx, leaving a recess on either side termed the *piriform fossa*. It is here that swallowed sharp foreign bodies such as fish bones tend to impact.

The internal branch of the superior laryngeal nerve passes in the submucosa of the piriform fossa. Local anaesthetic solutions applied to the surface of the piriform fossa on wool balls held in Krause's forceps will produce anaesthesia of the larynx above the vocal cords. This is a useful nerve block to supplement oral anaesthesia for laryngoscopy.

The structure of the pharynx

The pharynx has four coats: mucous, fibrous, muscular and fascial:
1 The mucosa is stratified squamous except in the nasopharynx, which is lined by a ciliated columnar epithelium. Beneath the surface are numerous mucous racemose glands.
2 The fibrous layer is relatively dense above (the *pharyngobasilar fascia*), where the muscle wall is deficient; it is also condensed to form the capsule of the tonsil and the posterior median raphe, but elsewhere it is thin.

3 The muscular coat is described below.

4 The fascial coat is the buccopharyngeal fascia, which is the very thin fibrous capsule of the pharynx.

CLINICAL NOTE

Ludwig's angina

Because of the fascial coat, inflammatory oedema may spread downwards from infections within the mouth or the tonsils or from dental sepsis. The spread of the oedema is restricted by the pharyngeal fascia and produces swelling and oedema of the tissues of the larynx and pharynx. This may produce difficulty in swallowing and then rapidly progresses to laryngeal obstruction unless the seriousness of the situation is realized and surgical drainage of the deep pharyngeal tissues performed. Similar complications can occur after operations involving the floor of the mouth. The anaesthetist should always consider the advisability of tracheostomy in these patients.

The muscles of the pharynx

The muscles of the pharynx are the superior, middle and inferior constrictors (which have been aptly likened to three flower-pots fitted into each other), the stylopharyngeus, salpingopharyngeus and palatopharyngeus.

The constrictor muscles (Fig. 15) have an extensive origin from the skull, mandible, hyoid and larynx on either side; they sweep round the pharynx to become inserted into the median raphe, which runs the length of the posterior aspect of the pharynx, being attached above to the pharyngeal

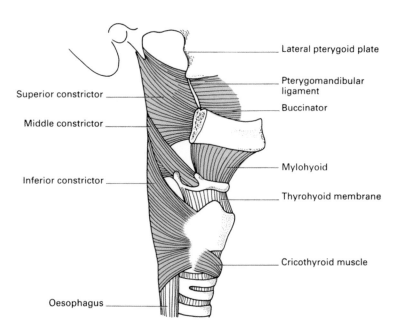

Fig. 15 The constrictor muscles of the pharynx.

Superior constrictor

Middle constrictor

Inferior constrictor

Oesophagus

Lateral pterygoid plate

Pterygomandibular ligament

Buccinator

Mylohyoid

Thyrohyoid membrane

Cricothyroid muscle

tubercle on the basilar part of the occipital bone and blending below with the oesophageal wall.

The *superior constrictor* muscle arises from the lower part of the medial pterygoid plate, the pterygoid hamulus, the pterygomandibular raphe and the posterior end of the mylohyoid line on the inner aspect of the mandible. The space between its upper free margin and the base of the skull allows the Eustachian tube to pass into the nasopharynx.

The *middle constrictor* spreads out like a fan from the lesser horn of the hyoid, the upper border of the greater horn and the lowermost part of the stylohyoid ligament.

The *inferior constrictor*, which is the thickest of the three, arises from the side of the cricoid, from the tendinous arch over the cricothyroid muscle and from the oblique line on the lamina of the thyroid cartilage. The muscle consists functionally of two parts: the lower portion, arising from the cricoid (the cricopharyngeus), acts as a sphincter, and its fibres are arranged transversely; the upper portion, with obliquely placed fibres that arise from the thyroid cartilage, has a propulsive action. Inco-ordination between these two components, so that the cricopharyngeus is in spasm while the thyropharyngeal element is initiating powerful peristalsis, is thought to be the aetiological basis for the development of a pharyngeal pouch. This first develops at a point of weakness posteriorly in the midline at the junction between the two portions of the muscle (Killian's dehiscence). As the pouch enlarges, it impinges first against the vertebral column, and then becomes deflected, usually to the more exposed left side (Fig. 16).

The constrictor muscles are supplied by the pharyngeal nerve plexus, which transmits the fibres of the accessory nerve in the pharyngeal branch of the vagus. In addition, the inferior constrictor receives filaments from the external branch of the superior laryngeal and the recurrent laryngeal branch of the vagus.

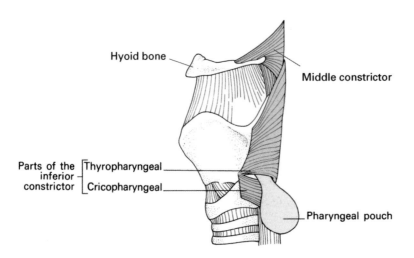

Hyoid bone

Middle constrictor

Parts of the inferior constrictor [Thyropharyngeal / Cricopharyngeal]

Pharyngeal pouch

Fig. 16 The pharyngeal pouch emerging between the two components of the inferior constrictor muscle.

Deglutition

The act of swallowing not only conveys food down the oesophagus but also disposes of mucus loaded with dust and bacteria from the respiratory passages. Moreover, during deglutition, the Eustachian auditory tube is opened, thus equalizing the pressure on either side of the ear drum.

Deglutition is a complex, orderly series of reflexes. It is initiated voluntarily but is completed by involuntary reflex actions set up by stimulation of the pharynx; if the pharynx is anaesthetized, normal swallowing cannot take place. The reflexes are co-ordinated by the deglutition centre in the medulla, which lies near the vagal nucleus and the respiratory centres.

The food is first crushed by mastication and lubricated by saliva; it is a common experience that it is well-nigh impossible to swallow a pill when the throat is dry. The bolus is then pushed back through the oropharyngeal isthmus by the pressure of the tongue against the palate, assisted by the muscles of the mouth floor.

During swallowing, the oral, nasal and laryngeal openings must be closed off to prevent regurgitation through them of food or fluid: each of these openings is guarded by a highly effective sphincter mechanism.

The nasopharynx is closed by elevation of the soft palate, which shuts against a contracted ridge of superior pharyngeal constrictor, the ridge of Passavant. At the same time, the tensor palati opens the ostium of the Eustachian tube. The oropharyngeal isthmus is blocked by contraction of palatoglossus on each side, which narrows the space between the anterior faucal pillars: the residual gap is closed by the dorsum of the tongue wedging into it.

The protection of the larynx is a complex affair, brought about not only by closure of the sphincter mechanism of the larynx but also by tucking the larynx behind the overhanging mass of the tongue and by utilizing the epiglottis to guide the bolus away from the laryngeal entrance. This mechanism may be interfered with as a result of an anterior flap tracheostomy (Bjork's tracheostomy). The consequent fixation of the trachea may limit mobility of the larynx and prevent its elevation during swallowing, resulting in aspiration of fluid into the trachea. The normal protective reflex is lost after the application of local anaesthetic solutions to the pharynx and after surgical interference with the pharyngeal muscles. The central nervous component of the swallowing reflex is depressed by opioids, anaesthesia and cerebral trauma. In these circumstances, aspiration of foreign material into the pulmonary tree becomes possible, particularly if the patient is lying on his/her back or in a head-up position.

The laryngeal sphincters are at three levels:

1 The aryepiglottic folds, defining the laryngeal inlet, which are apposed by the aryepiglottic and oblique interarytenoid muscles.
2 The walls of the vestibule of the larynx, which are approximated by the thyro-epiglottic muscles.
3 The vocal cords, which are closed by the lateral cricoarytenoid and transverse interarytenoid muscles.

The larynx is elevated and pulled forwards by the action of the thyrohyoid, stylohyoid, stylopharyngeus, digastric and mylohyoid muscles so that it comes into apposition with the base of the tongue, which is projecting backwards at this phase. While the larynx is raised and its entrance closed, there is reflex inhibition of respiration.

As the head of a bolus of food reaches the epiglottis, it is first tipped backwards against the pharyngeal wall and momentarily holds up the onward passage of the food. The larynx is then elevated and pulled forwards, drawing with it the epiglottis so that it now stands erect, guiding the food bolus into streams along both piriform fossae and away from the laryngeal orifice, like a rock that juts into a waterfall that will deviate the stream to either side. A little spill of fluid occurs into the laryngeal vestibule, often reaching as far as the false cords but seldom passing beyond them. Finally, the epiglottis flaps backwards as a cover over the laryngeal inlet, but this occurs only after the main bolus has passed beyond it. The epiglottis appears to act as a laryngeal lid at this stage to prevent deposition of fragments of food debris over the inlet of the larynx during re-establishment of the airway.

The cricopharyngeus then relaxes, allowing the bolus to cross the pharyngo-oesophageal junction. Fluids may shoot down the oesophagus passively under the initial impetus of the tongue action; semi-solid or solid material is carried down by peristalsis. The oesophageal transit time is about 15 seconds, relaxation of the cardia occurring just before the peristaltic wave reaches it. Gravity has little effect on the transit of the bolus, which occurs just as rapidly in the lying as in the erect position. It is, of course, quite easy to swallow fluid or solids while standing on one's head, a well-known party trick; here oesophageal transit is inevitably an active muscular process.

CLINICAL NOTE

The airway during anaesthesia

It is commonly perceived that, when a patient is anaesthetized in the supine position, the airway readily becomes obstructed as a result of the muscles of the jaw becoming relaxed and the tongue falling back to obstruct the oropharynx (Fig. 17a,b). This obstruction can be decreased by the use of an oropharyngeal airway. Studies have revealed that this may not be the complete explanation. Radiographs taken during induction of anaesthesia have shown that a more important cause of this obstruction is the blockage of the nasopharyngeal air passage brought about by the soft palate falling back onto the posterior nasopharyngeal mucosa. The sequence of events appears to be as follows:

1 the tongue obstructs the oral airway by impinging on the palate (hence snoring);
2 the nasal airway is blocked by the falling back of the soft palate.
   Relief of either of these obstructions will produce a clear airway.

The introduction of the laryngeal mask airway, inserted through the mouth to the laryngeal aperture, in which a cuff is inflated to produce a seal over the laryngeal orifice (Fig. 18), has provided an effective method of overcoming airway obstruction in the pharynx. Its use obviates the need to pull the tongue forwards manually in order to relieve the posterior pharyngeal obstruction in general anaesthesia.

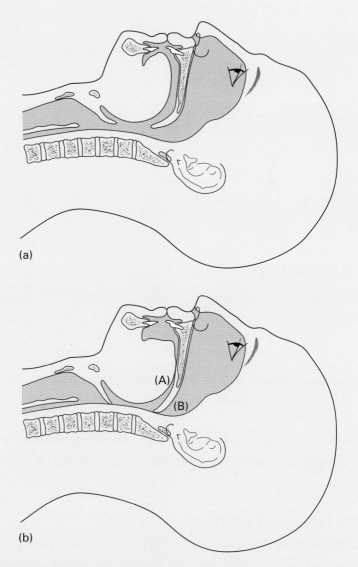

Fig. 17 (a) The relationship of the tongue to the posterior wall of the pharynx in the supine position in the conscious patient. (b) After induction of anaesthesia; both the tongue (A) and soft palate (B) move posteriorly.

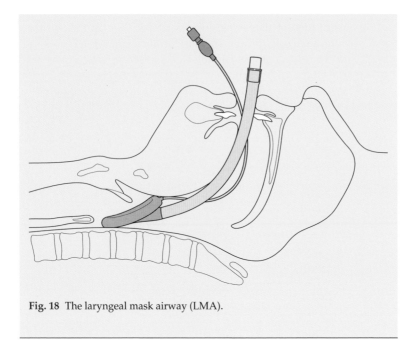

Fig. 18 The laryngeal mask airway (LMA).

The larynx

The competent anaesthetist should have a level of knowledge of the anatomy of the larynx of which a laryngologist would not be ashamed.

Evolutionally, the larynx is essentially a protective valve at the upper end of the respiratory passages to protect against inhalation of food during swallowing; its development into an organ of speech is a much later affair.

Structurally, the larynx consists of a framework of articulating cartilages, linked together by ligaments, that move in relation to each other by the action of the laryngeal muscles. It lies opposite the 4th, 5th and 6th cervical vertebrae (Fig. 19), separated from them by the laryngopharynx; its greater part is easily palpable, since it is covered superficially merely by the investing deep fascia in the midline and by the thin strap muscles laterally.

The laryngeal cartilages (Figs 20–23)

The principal cartilages are the thyroid, cricoid and the paired arytenoids, together with the epiglottis; in addition, there are the small corniculate and cuneiform cartilages.

The *thyroid cartilage* is shield-like and consists of two laminae that meet in the midline inferiorly, leaving the thyroid notch between them above. This junction is well marked in the male, forming the laryngeal prominence or Adam's apple, but in the female it is not obvious. The laminae

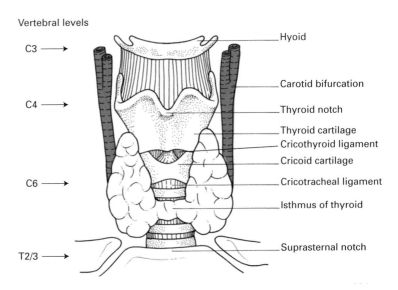

Vertebral levels

C3 →

C4 →

C6 →

T2/3 →

Hyoid

Carotid bifurcation

Thyroid notch

Thyroid cartilage
Cricothyroid ligament

Cricoid cartilage

Cricotracheal ligament

Isthmus of thyroid

Suprasternal notch

Fig. 19 Anterior view of the larynx to show adjacent important landmarks and vertebral levels.

carry superior and inferior horns, or cornua, at the upper and lower extremities of their posterior borders; the inferior horn bears a circular facet on its inner surface for the cricoid cartilage.

The *cricoid cartilage* is in the shape of a signet ring; the 'signet' lies posteriorly as a quadrilateral lamina joined in front by a thin arch. The side of the lamina bears two articular facets, one for the inferior horn of the thyroid cartilage and the other, near its upper extremity, for the arytenoid cartilage.

The *arytenoid cartilages* are three-sided pyramids that sit one on either side of the superolateral aspect of the lamina of the cricoid. Each has a lateral muscular process, into which are inserted the posterior and lateral cricoarytenoid muscles, and an anterior vocal process, which is the posterior attachment of the vocal ligament.

The *epiglottis* is likened to a leaf. It is attached at its lower tapering end to the back of the thyroid cartilage by means of the thyro-epiglottic ligament. Its superior extremity projects upwards and backwards behind the hyoid and the base of the tongue, and overhangs the inlet of the larynx. The posterior aspect of the epiglottis is free and bears a bulge, termed the *tubercle*, in its lower part. The upper part of the anterior aspect of the epiglottis is also free; its covering mucous membrane sweeps forwards centrally onto the tongue and, on either side, onto the side walls of the oropharynx, to form, respectively, the median glosso-epiglottic and the lateral glosso-epiglottic folds. The valleys on either side of the median glosso-epiglottic fold are termed the *valleculae*; they are common sites for the impaction of sharp swallowed objects such as fish bones.

The lower part of the anterior surface of the epiglottis is attached to the back of the hyoid bone by the hyo-epiglottic ligament. In the neonate, the epiglottis is more deeply furrowed at its free end, and in some babies it has a V-shaped appearance on laryngoscopy. The long, deeply grooved, 'floppy' epiglottis of the neonate more closely resembles that of aquatic mammals and is more suited to its function of protecting the nasotracheal air passage during suckling.

The *corniculate cartilage* is a small nodule lying at the apex of the arytenoid.

The *cuneiform cartilage* is a flake of cartilage within the margin of the aryepiglottic fold.

The laryngeal ligaments (Figs 20, 22–24)

The ligaments of the larynx can be divided into the extrinsic and the intrinsic, which link together the laryngeal cartilages.

The extrinsic ligaments are as follows.

1 The *thyrohyoid membrane*, which stretches between the upper border of the thyroid cartilage and the hyoid. This membrane is strengthened anteriorly by condensed fibrous tissue, termed the median thyrohyoid ligament, and its posterior margin is also thickened to form the lateral thyrohyoid ligament, stretched between the tips of the greater horn of the hyoid and the upper horn of the thyroid cartilage. The membrane is pierced by the internal branch of the superior laryngeal nerve and by the superior laryngeal vessels.

2 The *cricotracheal ligament*, which links the cricoid to the 1st ring of the trachea.

3 The *cricothyroid ligament* lies between the thyroid cartilage and the cricoid. It is an easily identified gap in the anterior surface of the laryngeal skeleton through which intratracheal injections may be administered. It is also the recommended site for emergency laryngotomy in cases of laryngeal obstruction (see below).

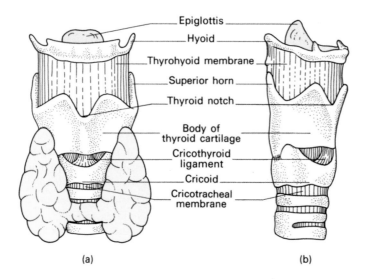

Epiglottis
Hyoid
Thyrohyoid membrane
Superior horn
Thyroid notch
Body of thyroid cartilage
Cricothyroid ligament
Cricoid
Cricotracheal membrane

(a)

(b)

Fig. 20 External views of the larynx: (a) anterior aspect; (b) anterolateral aspect with the thyroid gland and cricothyroid ligament removed.

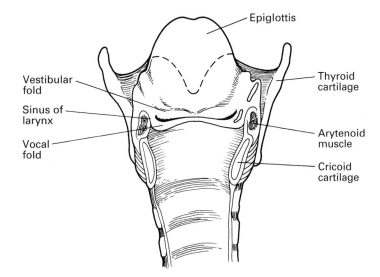

Epiglottis

Vestibular fold

Sinus of larynx

Vocal fold

Thyroid cartilage

Arytenoid muscle

Cricoid cartilage

Fig. 21 The larynx dissected from behind, with cricoid cartilage divided, to show the true and false vocal cords with the sinus of the larynx between.

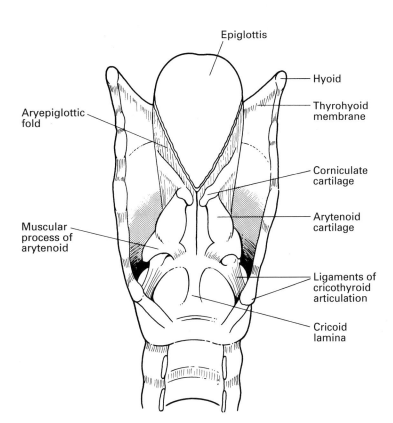

Epiglottis

Aryepiglottic fold

Muscular process of arytenoid

Hyoid

Thyrohyoid membrane

Corniculate cartilage

Arytenoid cartilage

Ligaments of cricothyroid articulation

Cricoid lamina

Fig. 22 The cartilages and ligaments of the larynx seen posteriorly.

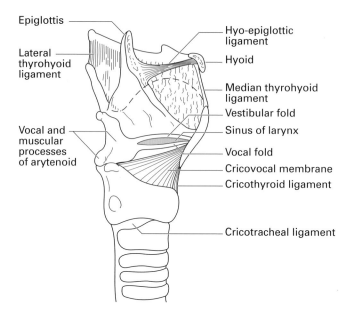

Epiglottis

Lateral thyrohyoid ligament

Vocal and muscular processes of arytenoid

Hyo-epiglottic ligament

Hyoid

Median thyrohyoid ligament

Vestibular fold

Sinus of larynx

Vocal fold

Cricovocal membrane

Cricothyroid ligament

Cricotracheal ligament

Fig. 23 The cartilages and ligaments of the larynx seen laterally.

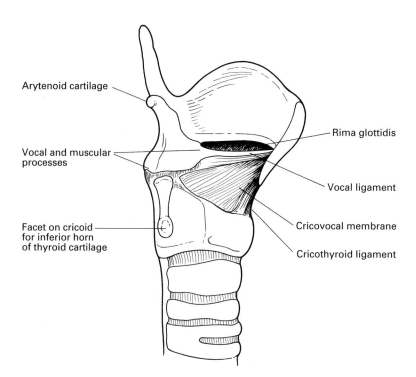

Arytenoid cartilage

Vocal and muscular processes

Facet on cricoid for inferior horn of thyroid cartilage

Rima glottidis

Vocal ligament

Cricovocal membrane

Cricothyroid ligament

Fig. 24 The formation of the vocal cord and cricovocal membrane.

4 The *hyo-epiglottic ligament*, which connects the epiglottis to the back of the body of the hyoid.

The intrinsic ligaments comprise the capsules of the tiny synovial joints between the arytenoid and cricoid, and between the thyroid and cricoid cartilages, which require no more than this passing mention; more important is the fibrous internal framework of the larynx.

If the cavity of the larynx is inspected in a bisected specimen, two folds will be seen, the upper vestibular and the lower vocal fold (or the *false* and *true vocal cords*), between which is a slit-like recess termed the *sinus* of the larynx (Fig. 21). From the anterior part of the sinus, the *saccule* of the larynx ascends as a pouch between the vestibular fold and the inner surface of the thyroid cartilage. Beneath the mucosa of the larynx is a sheet of fibrous tissue, divided into an upper and lower part by the sinus. The upper part of the sheet, termed the *quadrangular membrane*, forms the frame of the aryepiglottic fold, which is the fibrous skeleton of the laryngeal inlet; the lower margin of the quadrangular membrane is thickened to form the vestibular ligament, which underlies the vestibular fold, or false vocal cord. The lower sheet of fibrous tissue inferior to the sinus of the larynx contains many elastic fibres and forms the *cricovocal membrane* (Fig. 24). This is attached below to the upper border of the cricoid cartilage, and above is stretched between the mid-point of the laryngeal prominence of the thyroid cartilage anteriorly and the vocal process of the arytenoid behind. The free upper border of this membrane constitutes the *vocal ligament*, the framework of the true vocal cord. Anteriorly, the cricovocal membrane thickens into the cricothyroid ligament, which links the cricoid and thyroid in the midline.

CLINICAL NOTE

Tracheostomy, minitracheostomy and cricothyrotomy (cricothyroidotomy)

The performance of a formal tracheostomy requires time and a modicum of surgical skill. However, there are times when incipient or acute supraglottic airway obstruction occurs and an airway needs to be established. The 4th National Audit Project of the Royal College of Anaesthetists reported failure rates of up to 60% for cannula cricothyroidotomy (minitracheostomy) with an almost universal success for emergency surgical airway. It is now recommended that anaesthetists and intensivists should be trained to perform a surgical airway in the form of a dilatational tracheostomy.

Cricothyroidotomy and minitracheostomy

With the patient supine and the neck in the neutral position or, in the absence of cervical spine injury, in extension, the groove between the lower border of the thyroid cartilage and the cricoid cartilage is identified. This groove overlies the cricothyroid ligament. Depending upon the urgency and circumstances local anaesthetic is infiltrated subcutaneously, and a

1–2 cm horizontal incision is made over the cricothyroid ligament while the larynx is stabilized with the other hand. The ligament can then be incised in a 'stabbing' manner with a scalpel. The incision in the membrane can be enlarged by placing the handle of the scalpel into the hole and rotating the scalpel. A small tracheal tube or tracheostomy tube can then be passed through the incision, allowing ventilation of the lungs. Cricothyrotomy is relatively easy to perform and should (in theory at least) be associated with minimal blood loss, as the cricothyroid ligament is largely avascular (Fig. 25).

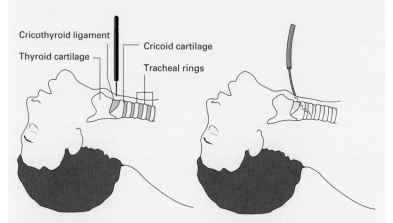

Cricothyroid ligament

Thyroid cartilage

Cricoid cartilage

Tracheal rings

Fig. 25 The site for cricothyroid puncture: the soft area between the thyroid cartilage and the cricoid is easily palpated and is relatively avascular.

Slightly less traumatic is the passage of a small diameter tracheostomy (minitracheostomy) tube using a Seldinger wire-guided technique or using an introducer. The positioning of the patient is the same as for cricothyrotomy but many operators favour a 1 cm vertical incision over the cricothyroid ligament. The membrane may be punctured using a needle and passing a wire, over which an introducer is inserted into the trachea. Alternatively, in some kits, the introducer itself is used to puncture the cricothyroid ligament. A 4.0 mm minitracheostomy tube is then inserted over the introducer as shown in Fig. 26. The minitracheostomy tube may be used for suction therapy for sputum retention, for O_2 therapy and as an emergency airway (for upper airway obstruction). In the UK, minitracheostomies are no longer used for ventilation.

Tracheostomy

The anterior relations of the cervical portion of the trachea are naturally of prime importance in performing a tracheostomy. It is important to keep the head fully extended with a sandbag placed between the patient's shoulders, and to maintain the head absolutely straight with the chin and sternal notch in a straight line. From the cosmetic point of view, it is better to use a short transverse incision placed midway between the cricoid cartilage and the suprasternal notch. The tyro may find in an emergency, however, that it is safer to use a vertical incision that passes from the lower border of the thyroid cartilage to just above the suprasternal notch.

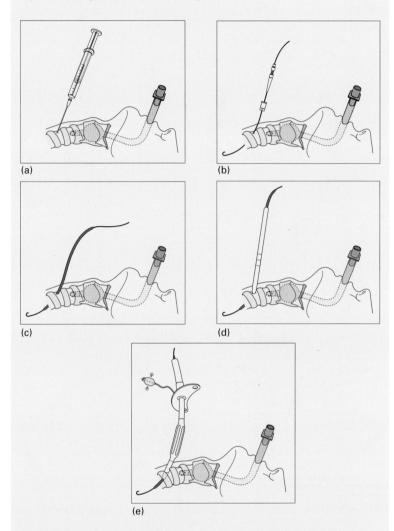

Fig. 26 Percutaneous tracheostomy. (a) Needle puncture below the first tracheal ring. (b) Insertion of a Seldinger wire. (c and d) Serial dilation of the aperture using bougies. (e) Insertion of the endotracheal tube.

The great anatomical and surgical secret of the operation is to keep exactly to the midline; in doing so, the major vessels of the neck are out of danger. The skin incision is deepened to the investing layer of fascia, which is split vertically, thus separating the pretracheal muscles on either side. These are held apart by retractors. The 1st ring of the trachea now comes into view, and the position of the trachea can be checked carefully by palpating its rings. It is often possible to push the isthmus of the thyroid gland downwards to expose the upper rings of the trachea; if not, the isthmus is lifted up by blunt dissection and divided vertically between artery forceps. The trachea is opened by a small vertical incision (Fig. 27). A tracheostomy tube of the largest size that will fit the tracheostome comfortably is inserted, the trachea is aspirated through it and the wound is loosely closed with two or three skin sutures.

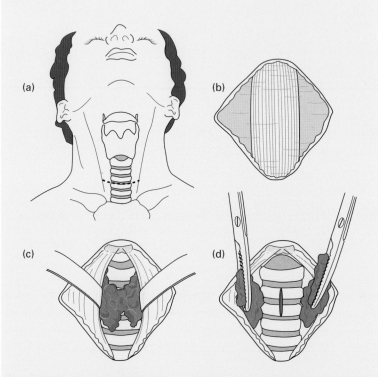

(a)

(b)

(c)

(d)

Fig. 27 Tracheostomy: (a) the incision is placed midway between the cricoid cartilage and the suprasternal notch. (b) The investing layer of fascia covering the pretracheal muscles is exposed. (c) The isthmus of the thyroid is cleared. This must either be divided between artery forceps or displaced downwards. (d) A vertical incision is made in the trachea.

Percutaneous tracheostomy

Percutaneous tracheostomy (Fig. 26) is now the tracheostomy method of choice on most intensive care units (ICUs). It may be performed by appropriately trained intensivists and is performed in the ICU rather than in the operating theatre. There are a variety of kits that work by progressive bougie dilatation or by balloon dilatation over a Seldinger wire inserted below the 1st tracheal ring under bronchoscopic control. The results of this new technique have been excellent and it is associated with fewer complications, in both the long and short term. The scarring is usually small and the cosmetic result satisfactory. Formal surgical tracheostomy is the preserve of the ENT surgeons and is reserved on the ICU for patients with difficult anatomy in whom there is a perceived risk of vascular damage or damage to the thyroid.

With the patient in the supine position and the neck extended, the cricoid cartilage is identified. The tracheal rings immediately below the cricoid cartilage are palpated. The space between the 1st and 2nd tracheal ring is most commonly used. Local anaesthetic, which may contain epinephrine in low concentration, is injected if necessary. A 1 cm incision is made over the trachea and a needle mounted on a syringe is passed through the tracheal wall between the cartilaginous rings. Aspiration of air confirms correct tracheal placement, and a wire is then passed through the needle into the trachea. The needle is withdrawn. This wire acts as a guide for a series of dilators, or a balloon dilator, that enlarge the hole in the trachea sufficiently to allow the passage of a tracheostomy tube. Great care must be exercised to ensure adequate ventilation and oxygenation during the performance of a percutaneous tracheostomy. This technique is currently accepted as the standard technique for longer term airway management in many ICUs.

The muscles of the larynx

The muscles of the larynx can be divided into the extrinsic group, which attach the larynx to its neighbours, and the intrinsic group, which are responsible for moving the cartilages of the larynx one against the other.

The *extrinsic muscles* of the larynx are the sternothyroid, thyrohyoid and the inferior constrictor of the pharynx. In addition, a few fibres of stylopharyngeus and palatopharyngeus reach forwards to the posterior border of the thyroid cartilage.

1 The *sternothyroid muscle* stretches from the posterior aspect of the manubrium to the oblique line on the lateral surface of the thyroid lamina. It is supplied by the ansa hypoglossi (see page 291) and depresses the larynx.

2 The *thyrohyoid muscle* passes upwards from the oblique line of the thyroid lamina to the inferior border of the greater horn of the hyoid. It is supplied by fibres of C1 conveyed through the hypoglossal nerve (see page 290). It elevates the larynx.

3 The *inferior constrictor* arises from the oblique line of the thyroid lamina, from a tendinous arch over the cricothyroid muscle and from the side of the pharynx. This muscle acts solely as a constrictor of the pharynx and is considered fully with this structure (see page 20).

Other muscles play an important part in movements of the larynx indirectly, via its close attachment, by ligaments and muscle, with the hyoid bone. These muscles help to elevate and depress the larynx; the indirect elevators are the mylohyoid, stylohyoid and geniohyoid, and the indirect depressors are the sternohyoid and omohyoid.

The *intrinsic muscles* of the larynx (Figs 28, 29) have a threefold function: they open the cords in inspiration, they close the cords and the

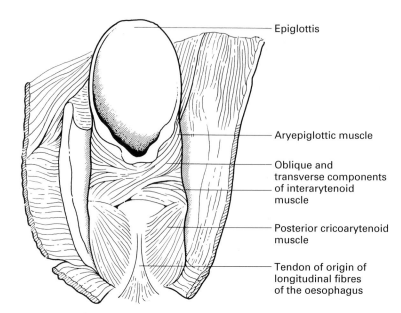

Epiglottis

Aryepiglottic muscle

Oblique and transverse components of interarytenoid muscle

Posterior cricoarytenoid muscle

Tendon of origin of longitudinal fibres of the oesophagus

Fig. 28 The intrinsic muscles of the larynx.

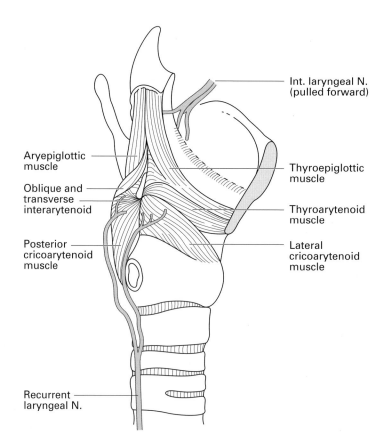

Int. laryngeal N. (pulled forward)

Aryepiglottic muscle

Oblique and transverse interarytenoid

Posterior cricoarytenoid muscle

Thyroepiglottic muscle

Thyroarytenoid muscle

Lateral cricoarytenoid muscle

Recurrent laryngeal N.

Fig. 29 The intrinsic muscles of the larynx, lateral view.

laryngeal inlet during deglutition, and they alter the tension of the cords during speech. They comprise the posterior and lateral cricoarytenoids, the interarytenoids and the aryepiglottic, the thyroarytenoid, the thyro-epiglottic, the vocalis and the cricothyroid muscles.

1 The *posterior cricoarytenoid* muscle arises from the posterior surface of the lamina of the cricoid and is inserted into the posterior aspect of the muscular process of the arytenoid. It abducts the cord by external rotation of the arytenoid and thus opens the glottis; it is the only muscle to do so.

2 The *lateral cricoarytenoid* muscle arises from the superior border of the arch of the cricoid and is inserted into the lateral aspect of the arytenoid cartilage. It adducts the cord by internally rotating the arytenoid cartilage, thus closing the glottis.

3 The *interarytenoid* muscle, the only unpaired muscle of the larynx, runs between the two arytenoid cartilages. Its action is to help close the glottis, particularly its posterior part. The muscle is made up of transverse and oblique fibres; the latter continue upwards and outwards as the *aryepiglottic* muscle, which lies within the aryepiglottic fold and acts as a rather feeble sphincter to the inlet of the larynx.

4 The *thyroarytenoid* muscle has its origin from the posterior aspect of the junction of the laminae of the thyroid cartilage and is inserted into the arytenoid cartilage on its anterolateral aspect, from the tip of its vocal process back onto its muscular process. By drawing the arytenoid forwards, this muscle serves to shorten, and thus relax, the vocal cord. Some fibres of this muscle continue in the aryepiglottic fold to the margin of the epiglottis, forming the *thyro-epiglottic* muscle, which assists in the sphincter mechanism of the laryngeal inlet.

5 The *vocalis* is simply some muscle fibres of the deep aspect of the thyroarytenoid that are inserted into the vocal fold. It may function as an adjusting mechanism to the tension of the cord.

6 The *cricothyroid*, the only intrinsic laryngeal muscle that lies outside the cartilaginous framework, arises from the anterior part of the outer aspect of the arch of the cricoid cartilage. Its fibres pass upwards and backwards to become inserted into the inferior border of the lamina of the thyroid cartilage and along the anterior face of its inferior cornu. Contraction of this muscle elevates the anterior part of the arch of the cricoid, approximating it to the thyroid cartilage. The effect of this is to tilt the lamina of the cricoid, bearing with it the arytenoid, posteriorly, thus lengthening the anteroposterior diameter of the glottis and thus, in turn, putting the vocal cords on stretch (Fig. 30). This muscle is the only tensor of the cord.

The actions of the intrinsic laryngeal muscles can be summarized thus:

1 abductors of the cords: posterior cricoarytenoids;
2 adductors of the cords: lateral cricoarytenoids, interarytenoid;
3 sphincters to vestibule: aryepiglottics, thyro-epiglottics;
4 regulators of cord tension: cricothyroids (tensors), thyroarytenoids (relaxors), vocales (fine adjustment).

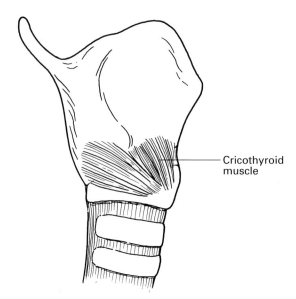

Cricothyroid
muscle

Fig. 30 The
cricothyroid muscle.

Blood supply

The *superior laryngeal artery* is a branch of the superior thyroid artery, which
is the 1st branch of the external carotid. It accompanies the internal branch
of the superior laryngeal nerve and in its company pierces the thyrohyoid
membrane to supply the interior of the larynx.

The *inferior laryngeal artery* arises from the inferior thyroid branch of the
thyrocervical trunk, which, in turn, arises from the first part of the subclavian artery. It accompanies the recurrent laryngeal nerve into the larynx. The corresponding veins drain into the superior and inferior thyroid
veins. Thus, the blood supply of the larynx comes from the superior and
inferior laryngeal arteries and veins, which are derived from the superior
and inferior thyroid vessels and which accompany the superior and 'inferior' (recurrent) laryngeal nerves, respectively.

Lymph drainage

The lymphatics of the larynx are separated into an upper and lower group
by the vocal cords. The supraglottic area is drained by vessels that accompany the superior laryngeal vein and empty into the upper deep cervical
lymph nodes. The infraglottic zone similarly drains in company with the
inferior vein into the lower part of the deep cervical chain. Lymphatics
from the anterior part of the lower larynx also empty into small prelaryn-
geal and pretracheal nodes.

The vocal cords themselves are firmly bound down to the underlying
vocal membrane; lymph channels are therefore absent in this region, which
accounts for the clearly defined watershed between the upper and lower
zones of lymph drainage.

Nerve supply

The nerve supply of the larynx is from the vagus via its superior and recurrent laryngeal branches.

The *superior laryngeal nerve* passes deep to both the internal and external carotid arteries and there divides into a small external branch, which supplies the cricothyroid muscle, and a larger internal branch, which pierces the thyrohyoid membrane to provide the sensory supply to the interior of the larynx as far down as the vocal cords; it probably also sends motor fibres to the interarytenoid muscle.

The *internal laryngeal nerve* runs beneath the mucosa of the piriform fossa. In this position it can easily be blocked by the topical application of local anaesthetic to provide anaesthesia for laryngoscopy and bronchoscopy.

The *recurrent laryngeal nerve* on the right side leaves the vagus as the latter crosses the right subclavian artery; it then loops under the artery and ascends to the larynx in the groove between the oesophagus and trachea. On the left side, the nerve originates from the vagus as it crosses the aortic arch; the nerve then passes under the arch to reach the groove between the oesophagus and the trachea. Once it reaches the neck, the left nerve assumes the same relationships as on the right (see Fig. 37). The recurrent laryngeal nerves provide the motor supply to the intrinsic muscles of the larynx apart from cricothyroid, as well as the sensory supply to the laryngeal mucosa inferior to the vocal cords.

Injuries of the laryngeal nerves

There is an intimate and important relationship between the nerves that supply the larynx and the vessels that supply the thyroid gland. The external branch of the superior laryngeal nerve descends over the inferior constrictor muscle of the pharynx immediately deep to the superior thyroid artery and vein as these pass to the superior pole of the gland; at this site the nerve may be damaged in securing these vessels. Since this nerve supplies the cricothyroid muscle, the sole tensor muscle of the vocal cord and hence described as the 'tuning fork of the larynx', its damage will be followed by hoarseness. This, fortunately, is temporary and becomes compensated by increased action of the opposite cricothyroid.

The recurrent laryngeal nerve, as it ascends in the tracheo-oesophageal groove, is overlapped by the lateral lobe of the thyroid gland, and here comes into close relationship with the inferior thyroid artery as this passes medially, behind the common carotid artery, to the gland. The artery may cross posteriorly or anteriorly to the nerve, or the nerve may pass between the terminal branches of the artery (Fig. 31). On the right side, there is an equal chance of locating the nerve in each of these three situations; on the left, the nerve is more likely to lie posterior to the artery. Injury to the recurrent nerve is an obvious hazard of thyroidectomy, especially since the nerve may be displaced from its normal anatomical location by a diseased thyroid gland.

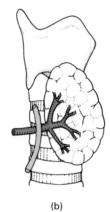

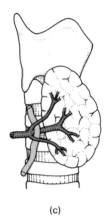

(a)   (b)   (c)

Fig. 31 The relationship between the recurrent laryngeal nerve and the inferior thyroid artery. The nerve may (a) cross posteriorly to the artery, (b) be anterior to or (c) pass through the branches of the artery. In this diagram, the lateral lobe of the thyroid is pulled forwards as it would be during thyroidectomy.

Recurrent laryngeal nerve paralysis may occur not only as a result of injury at thyroidectomy but also from involvement of the nerve by a malignant or occasionally benign enlargement of the thyroid gland, by enlarged lymph nodes or by cervical trauma. The left nerve may be involved in its thoracic course by malignant tumours of the lung or oesophagus, malignant or inflamed nodes, by an aneurysm of the aortic arch or even, in mitral stenosis, by compression between the left pulmonary artery (pushed upwards by the greatly enlarged left atrium) and the aortic arch. It is occasionally injured in performing a ligation of a patent ductus arteriosus, since the nerve lies immediately deep to the ductus as it hooks beneath the aortic arch (see Fig. 82a).

It is not surprising that the left recurrent nerve, whose intrathoracic course brings it into relationship with many additional structures, should be paralysed twice as often as the right. Some 25% of all recurrent nerve palsies, it should be noted, are idiopathic; they probably result from a peripheral neuritis.

Damage to the recurrent laryngeal nerve results in paralysis of the corresponding cord, which lies motionless, near the midline and at a lower level than the opposite side – the last being due to the downward drag of the paralysed muscles. Unilateral paralysis produces a slight hoarseness that usually disappears as a result of compensatory overadduction of the opposite normal cord. However, bilateral paralysis results in complete loss of vocal power. Moreover, the two paralysed cords flap together, producing a valve-like obstruction, especially during inspiration, with incapacitating dyspnoea and marked inspiratory stridor.

Respiratory obstruction after thyroidectomy can also result from direct trauma to the tracheal cartilages (especially in carcinoma of the thyroid) causing tracheomalacia. It may be the result of haemorrhage into the neck deep to the investing fascia, causing external pressure on the trachea. In practice, provided that the tracheal cartilages have not been damaged, it is very unusual for a benign enlarged thyroid gland to compress the trachea to an extent that prevents tracheal intubation. The trachea invariably straightens and dilates during intubation. Similarly, haemorrhage into an intact gland is more likely to obstruct the airway by producing laryngeal oedema than by compression of the intact trachea. However, following thyroidectomy, when the fascial planes have been disturbed, bleeding may compress the trachea and compromise the airway by both tracheal compression and laryngeal oedema.

It is of note that laryngoscopy within 24 hours of thyroidectomy often reveals some degree of oedema of the false cords, presumably as a result of external laryngeal trauma during the operation and damage to venous and lymphatic drainage channels.

CLINICAL NOTE

Laryngoscopic anatomy

To view the larynx at direct laryngoscopy and then to pass a tracheal tube depends on getting the mouth, the oropharynx and the larynx into one plane. Flexion of the neck brings the axes of the oropharynx and the larynx in line but the axis of the mouth still remains at right angles to the others; their alignment is achieved by full extension of the head at the atlanto-occipital joint. This is the position, with the nose craning forwards and upwards, that the anaesthetist assumes in sniffing the fresh air after a long day in the operating theatre, or in moving the head forwards to take the first sip from a pint of beer that is full to the brim.

At laryngoscopy, the anaesthetist first views the base of the tongue, the valleculae and the anterior surface of the epiglottis. The laryngeal aditus then comes into view (Fig. 32), bounded in front by the posterior aspect of the epiglottis, with its prominent epiglottic tubercle. The aryepiglottic folds are seen on either side running posteromedially from the lateral aspects of the epiglottis; they are thin in front but become thicker as they pass backwards where they contain the cuneiform and corniculate cartilages. The vocal cords appear as pale, glistening ribbons that extend from the angle of the thyroid cartilage backwards to the vocal processes of the arytenoids. Between the cords is the triangular (apex forwards) opening of the rima glottidis, through which can be seen the upper two or three rings of the trachea.

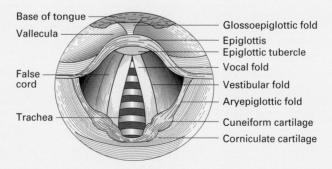

Fig. 32 View of the larynx at laryngoscopy.

Difficulties in tracheal intubation

Certain anatomical characteristics may make oral tracheal intubation difficult. This is particularly so in the patient with a poorly developed mandible
and receding chin, especially in those subjects in which this is associated
with a short distance between the angle of the jaw and the thyroid cartilage. A sagittal section through the head (Fig. 33) shows that the epiglottis
becomes 'tucked under' the bulging tongue, and great difficulty is experienced in such instances in inserting the blade of the laryngoscope into the
vallecula.

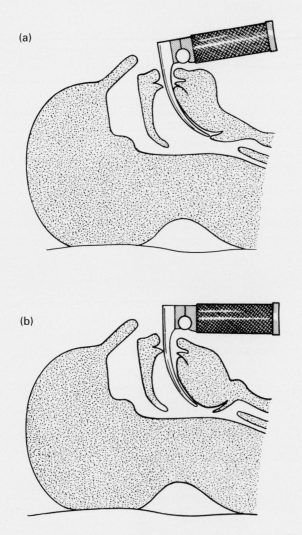

(a)

(b)

Fig. 33 (a) The position of the laryngoscope in the normal patient. (b) The
problem presented by the receding chin and poorly developed mandible.

Views of the larynx at laryngoscopy can be graded for purposes of assessment and record-keeping. The most popular grading system in current use is that described by Cormack and Lehane (Fig. 34).

A variety of scoring systems have been devised that aim to predict difficulties with laryngoscopy and intubation. The Mallampati score makes an anatomical assessment of the likely degree of difficulty of tracheal intubation based on the structures that can be seen when the patient opens his or her mouth fully (Fig. 35).

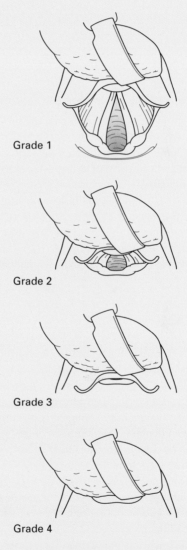

Grade 1

Grade 2

Grade 3

Grade 4

Fig. 34 The Cormack and Lehane laryngoscopy grading system. Grade 1, all structures are visible. Grade 2, only the posterior part of the glottis is visible. Grade 3, only the epiglottis is seen. Grade 4, no recognizable structures.

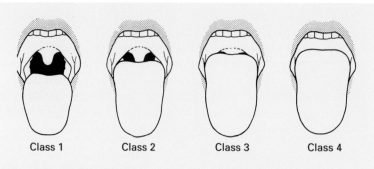

Class 1   Class 2   Class 3   Class 4

Fig. 35 The Mallampati scoring system. (After Mallampati 1985.)

Structure

The mucosa of the larynx is mainly a ciliated columnar epithelium with scattered mucus-secreting goblet cells. This epithelium becomes stratified squamous at two sites – over the vocal folds and at the entrance to the larynx where this is in continuity with the food passage, i.e. over the anterior aspect of the epiglottis, the upper part of its posterior aspect and the upper part of the aryepiglottic folds.

The mucosa also contains numerous mucous glands, particularly over the epiglottis. Here, these glands excavate the little pits that become obvious when the mucous membrane is stripped away from the epiglottic cartilage. The glands are absent, however, over the vocal folds, where the epithelium is adherent to the underlying vocal ligament.

The trachea

The trachea extends from its attachment to the lower end of the cricoid cartilage, at the level of the 6th cervical vertebra, to its termination at the bronchial bifurcation. In the preserved dissecting-room cadaver, this is at the level of the 4th thoracic vertebra and the manubriosternal junction (the angle of Louis); in the living subject in the erect position, the lower end of the trachea can be seen in oblique radiographs of the chest to extend to the level of the 5th, or in full inspiration the 6th, thoracic vertebra.

In the adult, the trachea is some 15 cm long, of which 5 cm lie above the suprasternal notch; this portion is somewhat greater (nearly 8 cm) when the neck is fully extended. The diameter of the trachea is correlated with the size of the subject; a good working rule is that it has the same diameter as the patient's index finger.

The patency of the trachea is maintained by a series of 16–20 C-shaped cartilages joined vertically by fibro-elastic tissue and closed posteriorly by the non-striated trachealis muscle. The cartilage at the tracheal bifurcation is the keel-shaped *carina* (Fig. 36), which is seen as a very obvious sagittal ridge when the trachea is inspected bronchoscopically. Should the sharp

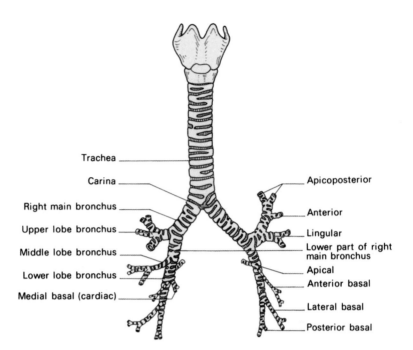

Trachea

Carina

Right main bronchus

Upper lobe bronchus

Middle lobe bronchus

Lower lobe bronchus

Medial basal (cardiac)

Apicoposterior

Anterior

Lingular

Lower part of right main bronchus

Apical

Anterior basal

Lateral basal

Posterior basal

Fig. 36 The trachea and main bronchi viewed from the front.

edge of the carina become flattened, this usually denotes enlargement of the hilar lymph nodes or gross distortion of the pulmonary anatomy by fibrosis, tumour or other pathology.

Relations

The trachea lies exactly in the midline in the cervical part of its course, but within the thorax it is deviated slightly to the right by the arch of the aorta.

In the *neck* (Fig. 37), it is covered anteriorly by the skin and by the superficial and deep fascia, through which the rings are easily felt. The 2nd to the 4th rings are covered by the isthmus of the thyroid where, along the upper border, branches of the superior thyroid artery join from either side. In the lower part of the neck, the edges of the sternohyoid and sternothyroid muscles overlap the trachea, which here is also covered by the inferior thyroid veins (as they stream downwards to the brachiocephalic veins), by the cross-communication between the anterior jugular veins and, when present, by the thyroidea ima artery, which ascends from the arch of the aorta or from the brachiocephalic artery. It is because of this close relationship with the brachiocephalic artery that erosion of the tracheal wall by a tracheostomy tube may cause sudden profuse haemorrhage. It is less common for the carotid artery to be involved in this way. On either side are the lateral lobes of the thyroid gland, which intervene between the trachea and the carotid sheath and its contents (the common carotid artery, the internal jugular vein and the vagus nerve). Posteriorly, the trachea rests on the oesophagus, with the recurrent laryngeal nerves lying on either side in a groove between the two.

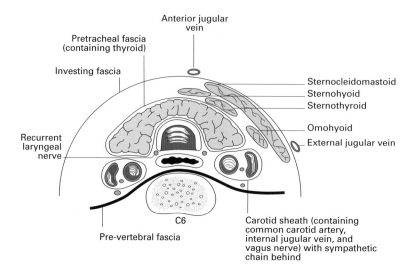

Fig. 37 Transverse section of the neck through C6.

The close relationship of the unsupported posterior tracheal wall and the oesophagus is revealed during oesophagoscopy. When a tracheal tube with an inflated cuff is in the trachea, the anterior wall of the oesophagus is compressed. For this reason, patients with inflated tracheostomy tubes (especially high-pressure cuffs) may have difficulty in swallowing. During oesophagoscopy with a rigid oesophagoscope, an overinflated tracheal tube cuff may be mistaken for an oesophageal obstruction.

Because the trachea is a superficial structure in the neck, it is possible to feel the bulge caused by the rapid injection of 5 ml of air into the cuff of an accurately placed tracheal tube. This is detected by placing two fingers over the trachea above the suprasternal notch.

The *thoracic* part of the trachea (Fig. 38) descends through the superior mediastinum. Anteriorly, from above downwards, lie the inferior thyroid veins, the origins of the sternothyroid muscles from the back of the manubrium, the remains of the thymus, the brachiocephalic artery and the left common carotid artery – which separate the trachea from the left brachiocephalic vein – and, lastly, the arch of the aorta. Posteriorly, as in its cervical course, the trachea lies throughout on the oesophagus, with the left recurrent laryngeal nerve placed in a groove between the left borders of these two structures (Fig. 39).

On the right side, the trachea is in contact with the mediastinal pleura, except where it is separated by the azygos vein and the right vagus nerve. On the left, the left common carotid and left subclavian arteries, the aortic arch and the left vagus intervene between the trachea and the pleura; the altering relationships between the major arteries and the trachea are the result of the diverging, somewhat spiral, course of the arteries from their aortic origins to the root of the neck.

The large tracheobronchial lymph nodes lie at the sides of the trachea and in the angle between the two bronchi.

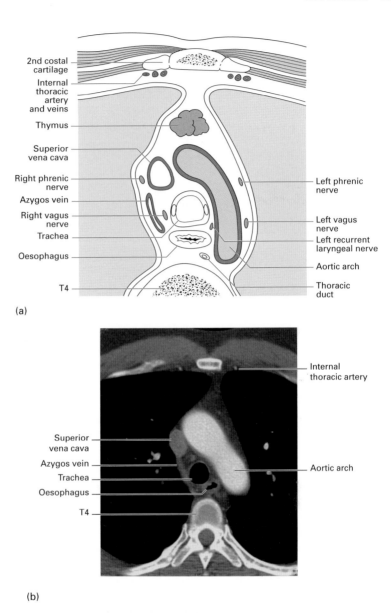

Fig. 38 The thoracic trachea and its environs in a transverse section through the 4th thoracic vertebra. (a) Diagrammatic. (b) Computerized tomography at T4.

In *infants*, these relationships are somewhat modified; the brachio-cephalic artery is higher and crosses the trachea just as it descends behind the suprasternal notch. The left brachiocephalic vein may project upwards into the neck to form an anterior relation of the cervical trachea – a frightening encounter if found tensely distended with blood when performing a tracheotomy on an asphyxiating baby. In children up to the age of 2 years, the thymus is large and lies in front of the lower part of the cervical trachea.

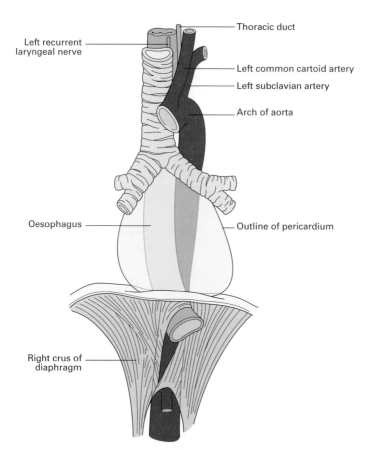

Thoracic duct

Left recurrent laryngeal nerve

Left common cartoid artery

Left subclavian artery

Arch of aorta

Oesophagus

Outline of pericardium

Right crus of diaphragm

Fig. 39 The trachea and its immediate relations.

Vascular, lymphatic and nerve supply

The arterial supply to the trachea is derived from the inferior thyroid arteries and the venous drainage is via the inferior thyroid veins. Lymphatics pass to the deep cervical, pretracheal and paratracheal nodes. The trachea is innervated by the recurrent laryngeal branch of the vagus with a sympathetic supply from the middle cervical ganglion.

The main bronchi

The trachea bifurcates in the supine cadaver at the level of the 4th thoracic vertebra into the right and left bronchi. In the erect position in full inspiration in life, the level of bifurcation is at T6.

The *right main bronchus* (Fig. 36) is shorter, wider and more vertically placed than the left: shorter because it gives off its upper lobe bronchus sooner (after a course of only 2.5 cm); wider because it supplies the larger lung; and more vertically placed (at 25° to the vertical compared with 45° on the left) because the left bronchus has to extend laterally behind the aortic arch to reach its lung hilum. Obviously, inhaled foreign bodies or a bronchial aspirating catheter are far more inclined to enter the wider and more vertical right bronchus than the narrower and more obliquely placed left.

The right pulmonary artery is first below and then in front of the right main bronchus, and the azygos vein arches over it.

The *left main bronchus* (Figs 36, 39) is 5 cm long. It passes under the aortic arch, in front of the oesophagus, thoracic duct and descending aorta, and has the left pulmonary artery lying first above and then in front of it.

CLINICAL NOTE

Endobronchial tubes

Because the right upper lobe bronchus arises only a short distance below the carina, it is not possible to place a tube in that bronchus without the risk of obstruction of the lower lobe. To overcome this difficulty, right-sided endobronchial tubes have an orifice in the lateral surface of the tube that coincides with the opening of the right upper lobe (Fig. 40). No special arrangement has to be made for tubes placed in the left bronchus, as the 5 cm distance between the carina and the left upper lobe bronchus leaves ample room for the cuffed end of an endobronchial tube. The position of the tube is routinely checked by fibre-optic bronchoscopy.

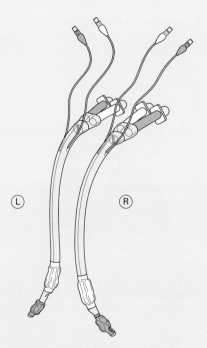

Fig. 40 Left and right double-lumen endobronchial tubes. Note the fenestration in the area of the distal cuff on the right-side tube.

The pleura

Each lung, in its development, invaginates the coelomic cavity to form a double-walled serous-lined sac – the pleura. The pleura comprise a *visceral layer*, which invests the lung itself and passes into its fissures, and a *parietal layer*, which clothes the diaphragm, the chest wall, the apex of the thorax and the mediastinum. The two layers meet at the site of invagination, which is the lung root or *hilum*, and here the pleura hang down as a fold, rather like an empty sleeve, termed the *pulmonary ligament*. Between the two layers of the pleura is a potential space, the *pleural cavity*, which is moistened with a film of serous fluid.

The visceral pleura is closely adherent to the surface of the underlying lung; attempts to strip the pleura will tear into the lung parenchyma. In contrast, the parietal pleura is separated from the overlying chest wall by a thin layer of loose connective tissue, the *extrapleural fascia*, which enables the surgeon to strip the parietal pleura freely and bloodlessly from the chest wall.

The lines of pleural reflection (Figs 41, 42)

The pleural margins can be mapped out on the chest wall as follows.
1 The apex of the pleura extends about 4 cm above the medial third of the clavicle.
2 The margin then passes behind the sternoclavicular joint and meets the opposite pleural edge behind the sternum at the 2nd costal cartilage level (the angle of Louis).
3 At the 4th cartilage, the left pleura deflects to the lateral margin of the sternum, corresponding to the cardiac notch of the underlying lung, and descends thence to the 6th costal cartilage.

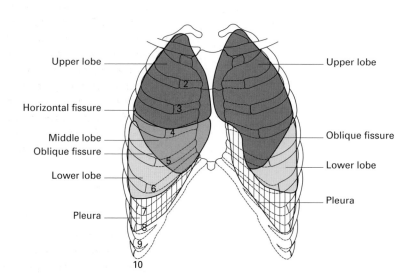

Fig. 41 The surface markings of the lungs and pleura, anterior view.

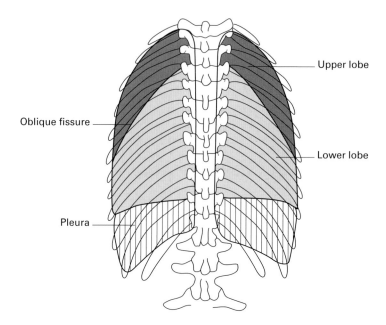

Upper lobe

Oblique fissure

Lower lobe

Pleura

Fig. 42 The surface markings of the lungs and pleura, posterior view.

4 The right pleural edge continues vertically downwards and projects a little below the right costo-xiphoid angle.

5 The pleural lower margin is easily remembered; it lies at the level of the 8th rib in the mid-clavicular line and at the level of the 10th rib at the mid-axillary line (here at its lowest level), and terminates behind at the level of the spine of the 12th thoracic vertebra, descending posteriorly slightly below the costal margin at the costo-vertebral angle.

The inferior margin of the parietal pleura does not extend to the line of attachment of the diaphragm to the chest wall; here, the muscle of the diaphragm comes into direct contact with the lowermost costal cartilages and intercostal spaces. Moreover, the lung in quiet respiration does not fill the lowermost extremity of the pleural sac, but leaves the slit-like *costo-diaphragmatic recess* where the costal and diaphragmatic pleurae are in contact. This recess provides a reserve sinus which the lung invades in forced inspiration; it may be opened inadvertently in operations performed below, or through the bed of, the 12th rib, notably during exposure of the kidney or the suprarenal gland.

CLINICAL NOTE

Chest tube insertion (Fig. 43)
Drainage of air or liquids such as blood from the pleura often requires the placement of a chest drain/tube. The British Thoracic Society recommends that, when inserting a drain for fluid drainage, ultrasonic guidance be used to identify loculations and pleural thickening. Real-time scanning improves safety as the diaphragm can be identified during respiration. In extreme

circumstances, this may not be possible or feasible. Insertion should be in the safe triangle as shown in Fig. 44. This is the triangle bordered by the anterior border of latissimus dorsi, the lateral border of the pectoralis major muscle, a horizontal line corresponding to the 5th intercostal space and an apex below the axilla. The hand of the affected side is placed behind the head to expose the area. After identification of the correct site of placement, local anaesthetic (lidocaine with epinephrine up to 3 mg/kg) is infiltrated subcutaneously if necessary and a small incision is made in the skin.

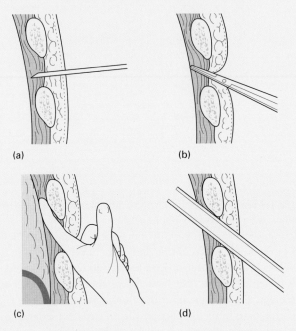

(a)   (b)

(c)   (d)

Fig. 43 Chest drain insertion. (a) Local anaesthetic is infiltrated into an intercostal space. (b) After an incision is made, blunt dissection allows access to the pleura. (c) A finger is passed through the incision to clear the lung away. (d) A chest tube is passed through the incision into the chest.

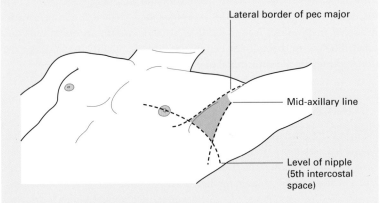

Lateral border of pec major

Mid-axillary line

Level of nipple (5th intercostal space)

Fig. 44 Safe triangle for chest drain insertion.

Small-bore tubes (8–14F) may be inserted without blunt dissection. A needle and syringe are used to identify air/fluid and a guidewire inserted down the needle. A dilator is used to dilate the tract and a small-bore tube may be passed into the cavity under imaging guidance. Medium-sized tubes (16–24F) can be inserted using either a Seldinger technique or blunt dissection.

Larger tubes require incision and blunt dissection using a Spencer Wells forceps or similar instrument over the upper edge of a rib. In this way, the intercostal nerves and vessels that pass immediately inferior to the rib above are avoided (Fig. 45). A finger can be passed into the pleural space to ensure that no lung adhesions are in the vicinity, and then a chest tube (without a sharpened trocar if supplied) can be grasped with large forceps and passed through the incision into the pleural cavity.

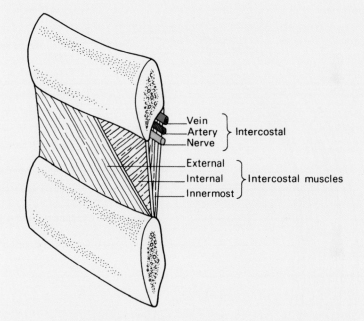

Fig. 45 The composition of an intercostal space.

The position of all chest/intrapleural catheters should be confirmed using a chest radiograph following insertion.

The intercostal spaces

The intercostal spaces are closed by thin but strong muscles and aponeuroses between which course the nerves, blood vessels and lymphatics of the chest wall (Fig. 45).

The intercostal muscles

The muscles of the intercostal spaces are disposed in three layers corresponding to the three layers of the lateral abdominal wall (Fig. 46).

The *external intercostals* pass downwards and forwards from the lower border of one rib to the upper border of the rib below, and extend from the tubercle of the rib posteriorly to the neighbourhood of the costochondral junction in front. Anteriorly, each is continued as the tough *anterior intercostal membrane* to the side of the sternum.

The *internal intercostals* extend from the sternum (or the anterior extremity of the costal cartilages in the lower spaces) to the angle of the rib posteriorly; thence, each is replaced by the *posterior intercostal membrane*. The muscle fibres run obliquely downwards and backwards.

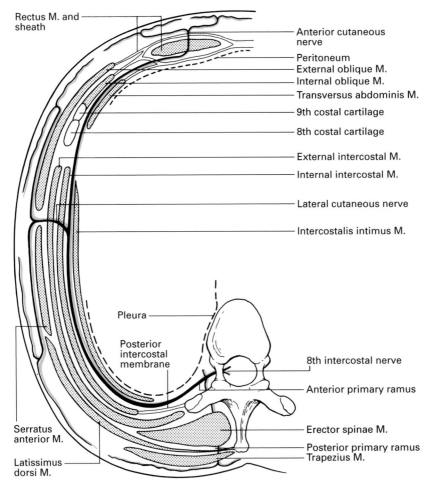

Fig. 46 Section of the distribution of the 8th intercostal nerve to show its relationships in both its thoracic and abdominal course to the muscle layers of the body wall.

The *innermost layer* of muscles is incomplete and comprises the following:

1 *Transversus thoracis* (*sternocostalis*) anteriorly, which fans out from the side of the lower sternum to the costal cartilages of the 2nd to 6th ribs.
2 The *intracostals* (*intercostales intimi*) laterally, which blend with the internal intercostals except where separated by the neurovascular bundle.
3 The *subcostals* posteriorly, made up of small slips near the angles of the lower ribs. They run in the same direction as the internal intercostals but span two or three interspaces.

These muscle slips of the internal layer are linked to each other by membranous tissue that is continuous superiorly with the suprapleural membrane (Sibson's fascia).

The *endothoracic fascia*, equivalent to the transversalis fascia of the abdominal wall, is no more than a fine layer of areolar connective tissue between the muscles of the intercostal spaces and the parietal pleura. This allows the surgeon to strip the parietal pleura from the chest wall with ease.

The intercostal muscles in respiration

Paralysis of the intercostal muscles is known to decrease the amplitude of rib movements during quiet respiration, but the exact role of the intercostal muscles has, however, been the subject of much controversy; animal experiments, models and geometry have been invoked in attempts to solve this problem. Electromyography has greatly helped to elucidate the action of these muscles. During *inspiration*, the intercostals contract and their electrical activity increases with greater respiratory effort. During this phase there is elevation and eversion of the ribs with increase in the anteroposterior and lateral diameters of the thorax. The intercostal muscles of the lower rib spaces also contract during forcible *expiration*. In this phase, the muscles probably act from their insertions; the lower ribs are fixed by contraction of the abdominal muscles and the intercostal muscles of the lower spaces then draw down the ribs and help to diminish the thoracic volume. In addition, the intercostal muscles by their contraction maintain the rigidity of the rib spaces during violent expiration. Rigid fixation of the chest wall, for example in ankylosing spondylitis, reduces the maximum breathing capacity of the patient by some 20–30%.

The neurovascular bundle

In each intercostal space lies a neurovascular bundle comprising, from above downwards, the posterior intercostal vein, the posterior intercostal artery and the intercostal nerve, protected by the costal groove of the upper rib (Fig. 45). Posteriorly, this bundle lies between the pleura and the posterior intercostal membrane, but at the angle of the rib it passes between the internal intercostal and the intracostal muscles.

The blood vessels of the chest wall

The *posterior intercostal arteries* of spaces 3–11 arise directly from the thoracic aorta. Those of the 1st and 2nd spaces are derived from the superior intercostal artery, a branch of the costo-cervical trunk, which arches over Sibson's fascia and crosses the front of the neck of the 1st rib. In this position, the sympathetic trunk is medial and the 1st thoracic nerve root is lateral to the artery.

Each posterior intercostal artery gives off a collateral branch, and these two vessels anastomose with the two anterior intercostal arteries (see below) in the upper nine spaces.

The *posterior intercostal veins*, lying above their corresponding arteries, have a somewhat complex and variable termination.

The vein of the 1st space on each side drains into either the vertebral or the brachiocephalic vein. The 2nd, 3rd and sometimes the 4th vein join to form the *superior intercostal vein*. On the right, this enters the azygos vein; on the left, it runs across the arch of the aorta, here passing between the phrenic and vagus nerves, to terminate in the left brachiocephalic vein. The lowest eight veins empty on the right into the azygos vein and on the left into the superior and inferior hemiazygos veins (four into each).

The *internal thoracic artery* arises from the first part of the subclavian artery, descends behind the upper six costal cartilages a finger's breadth from the lateral border of the sternum, and terminates by dividing into the superior epigastric and the musculophrenic arteries. At first, the artery lies directly against the pleura but then, at the 3rd costal cartilage, it passes in front of transversus thoracis, which protects it until its termination.

Perforating branches pierce each intercostal space to supply overlying pectoralis major, skin and, in the female, breast (the 2nd to 4th being the largest vessels to this gland). Other vessels supply the mediastinum and pericardium and one accompanies the phrenic nerve (the pericardiacophrenic artery). In each of the upper nine intercostal spaces, two anterior intercostal arteries are given off that anastomose with the posterior intercostal artery and its collateral branch; the first six of these branches derive from the internal thoracic artery, those to the 7th to 9th spaces arise from its musculophrenic branch.

The *internal thoracic vein* accompanies the artery and drains into the corresponding brachiocephalic vein.

Lymphatics

Lymph nodes lie alongside the internal thoracic vessels; these are important because they receive lymph from the breast which is discharged thence into the thoracic duct or mediastinal lymph trunk.

Nerves of the chest wall

The *intercostal nerves* are the anterior primary rami of T1–11; each lies in the neurovascular bundle already described. The lower five nerves (T7–11) continue onwards to supply the abdominal wall, as described elsewhere (page 319), where they maintain their anatomical position between the 2nd and 3rd layers of muscle of the body wall, i.e. between internal oblique and transversus abdominis (Fig. 46). These nerves form the segmental sensory nerve supply to the trunk and medial side of the upper arm, and the motor supply to the intercostal and anterior abdominal wall muscles.

The 1st intercostal nerve is atypical; it is the largest of the thoracic rami because of its branch which crosses the neck of the 1st rib to join C8 in the formation of the lowest trunk of the brachial plexus. Its intercostal branch is small and is entirely motor.

Each of the remaining intercostal nerves has the following branches.

1 *Collateral*, which arises at the angle of the rib and ends either in supplying muscle or as a connecting loop with the main nerve; it is entirely motor.

2 *Lateral cutaneous*, which arises in the mid-axillary line and gives off an anterior and posterior branch.

3 *Anterior cutaneous*, which, in each of the upper six intercostal spaces, passes in front of the internal thoracic vessels, then surfaces to supply the overlying skin. The lower five nerves pierce rectus abdominis to supply the anterior abdominal wall (Fig. 47).

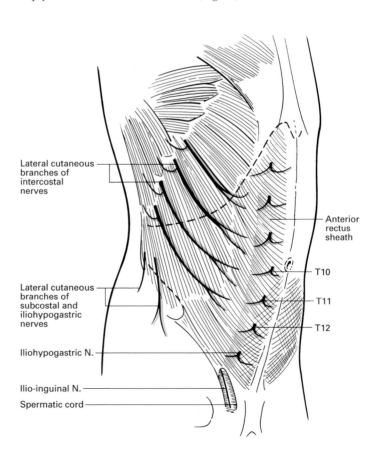

Lateral cutaneous branches of intercostal nerves

Anterior rectus sheath

T10

T11

T12

Lateral cutaneous branches of subcostal and iliohypogastric nerves

Iliohypogastric N.

Ilio-inguinal N.

Spermatic cord

Fig. 47 View of the abdominal wall after removing the skin and superficial fascia. The lateral cutaneous and anterior cutaneous terminal branches of the nerves of the abdominal wall are demonstrated.

The lateral cutaneous branch of the 2nd intercostal is atypical; it forms the *intercostobrachial* nerve that arches over the roof of the axilla to supply the skin of the medial aspect of the upper arm as far as the elbow. As it is not part of the brachial plexus, it is not affected by brachial plexus blocks, a point of importance when an upper arm tourniquet is used on the awake patient. Local anaesthetic needs to be deposited subcutaneously at the axilla along the medial border of the upper arm in order to provide analgesia for the tourniquet.

CLINICAL NOTE

Intercostal nerve block (Fig. 48)
Intercostal nerve blocks can be used for postoperative analgesia after thoracic or upper abdominal surgery or to treat the pain of fractured ribs. They may also be used in chronic pain conditions such as post-herpetic neuralgia. The blocks should be performed at approximately the angle of the rib, with the patient in the sitting, lateral or prone position. After subcutaneous infiltration with local anaesthetic, the lower border of a rib is palpated and the palpating finger moves the skin slightly cephalad. A short hypodermic needle is inserted onto the rib towards its lower border. The needle is then 'walked' off the lower edge of the rib, and then it is advanced 3–5 mm in a slightly cephalad direction. After careful aspiration to exclude intravascular placement, 3–5 ml of local anaesthetic is injected. Care must be exercised with regard to the total dose of local anaesthetic drug injected, as multiple injections are often needed and absorption of the local anaesthetic into the circulation from this site is rapid.

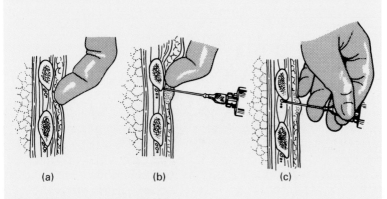

| (a) | (b) | (c) |

Fig. 48 Technique of intercostal blocks.

The mediastinum

The mediastinum is the region between the two pleural sacs. It is divided by the pericardium, somewhat artificially, into four compartments, which

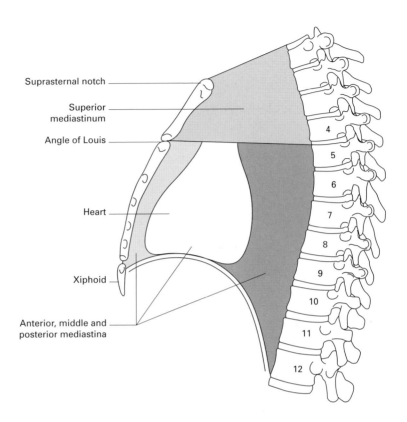

Suprasternal notch

Superior mediastinum

Angle of Louis

Heart

Xiphoid

Anterior, middle and posterior mediastina

4
5
6
7
8
9
10
11
12

Fig. 49 The subdivisions of the mediastinum.

are, however, useful for descriptive purposes (Fig. 49): the *middle medi-astinum* is the space occupied by the pericardium and its contents; the *anterior mediastinum* lies between this and the sternum; the *posterior medi-astinum* lies behind the pericardium above and the diaphragm below; and the *superior mediastinum* is situated between the pericardium and the thoracic inlet.

The lungs

The shape of the lungs is a reflection of the shape of the pleural cavity on either side, very much as a jelly is a reflection of the shape of its mould. Each lung is roughly conical, with an apex, a base, a lateral (or costal) and a medial surface and with three borders – anterior, posterior and inferior (Figs 50, 51). Each lung lies freely within its pleural cavity apart from its attachments at the hilum. The right lung is the larger, weighing on average 620 g compared with 570 g on the left. The lung of the male is larger and heavier than that of the female.

The *apex* of the lung extends upwards into the root of the neck, its sum-mit reaching 4 cm above the medial third of the clavicle. At this site, the pleura are in danger of puncture when a supraclavicular brachial plexus

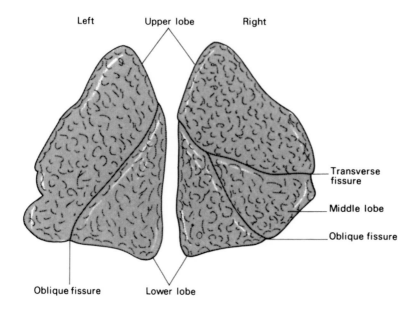

Fig. 50 The right and left lungs, posterolateral aspects.

block is performed, in subclavian vein puncture and in stab wounds of the neck. Because of the obliquity of the thoracic inlet (Fig. 49), the apex does not rise posteriorly above the neck of the 1st rib. The apex is grooved by the subclavian artery from which it is separated by cervical pleura and by the extrapleural (Sibson's) fascia.

The concave *base* of the lung rests on the dome of the diaphragm; since the right diaphragm is higher than the left, the right lung, although larger, is squatter than the left. The costal surface relates to the ribcage, which

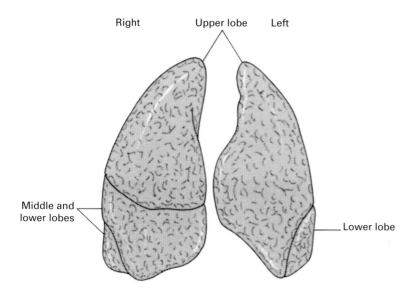

Fig. 51 The right and left lungs, anterior aspects.

produces the grooving seen on the formalin-hardened lung of the dissecting-room cadaver.

The medial surface has the *hilum* as its most striking feature (see page 62). It is also related to the structures within the mediastinum, the more prominent of which produce impressions on the hardened lung; this surface may be divided into a posterior (or vertebral) part, which is in contact with the thoracic vertebral column, and an anterior (or mediastinal) portion. This is deeply concave immediately below and in front of the lung hilum, forming the cardiac impression of the heart – naturally this impression is deeper on the left than on the right (Figs 52, 53).

The medial relationships of the two lungs are different. On the right mediastinal surface (Fig. 54), the cardiac impression is formed by the right atrium and part of the right ventricle. In addition, this surface is in contact with the superior and inferior venae cavae, the vena azygos (as this arches over the hilum), the right margin of the oesophagus, the trachea, the right vagus and phrenic nerves.

The left mediastinal surface is related to the left auricle and ventricle, the aortic arch and the descending aorta, the left subclavian and common carotid arteries, left brachiocephalic vein, trachea and oesophagus, the left vagus and the thoracic duct (Fig. 55).

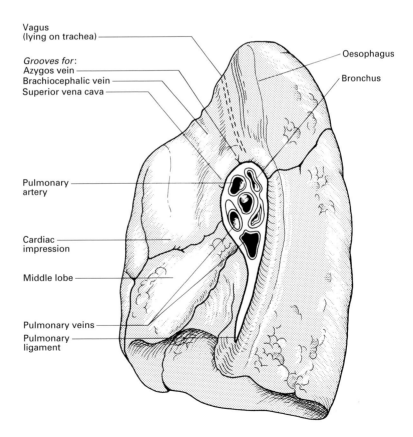

Vagus
(lying on trachea)

Grooves for:
Azygos vein
Brachiocephalic vein
Superior vena cava

Oesophagus

Bronchus

Pulmonary
artery

Cardiac
impression

Middle lobe

Pulmonary veins
Pulmonary
ligament

Fig. 52 The mediastinal aspect of the right lung.

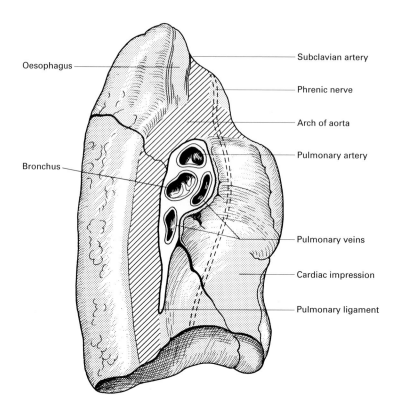

Oesophagus

Bronchus

Subclavian artery

Phrenic nerve

Arch of aorta

Pulmonary artery

Pulmonary veins

Cardiac impression

Pulmonary ligament

Fig. 53 The mediastinal aspect of the left lung.

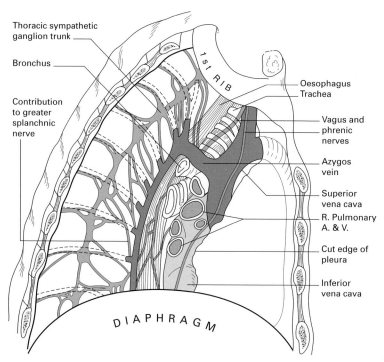

Thoracic sympathetic ganglion trunk

Bronchus

Contribution to greater splanchnic nerve

1st RIB

Oesophagus
Trachea

Vagus and phrenic nerves

Azygos vein

Superior vena cava

R. Pulmonary A. & V.

Cut edge of pleura

Inferior vena cava

DIAPHRAGM

Fig. 54 The right mediastinum: the medial relationships of the right lung.

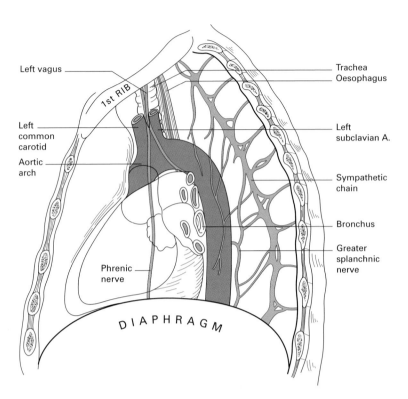

Left vagus

1st RIB

Trachea
Oesophagus

Left
common
carotid

Left
subclavian A.

Aortic
arch

Sympathetic
chain

Bronchus

Greater
splanchnic
nerve

Phrenic
nerve

DIAPHRAGM

Fig. 55 The left mediastinum: the medial relationships of the left lung.

The anterior borders of the lungs are thin and insinuate themselves between the pericardium and the chest wall. On the left side, this border bears a prominent *cardiac notch*; this leaves an area of right ventricle to come into contact with the chest wall, the pericardium only intervening.

The *surface projection* of the lung is somewhat less extensive than that of the parietal pleura (Figs 41, 42) and, in addition, it varies quite considerably with the phase of respiration. The apex of the lung closely follows the line of the cervical pleura, and the surface marking of the anterior border of the right lung corresponds to that of the right mediastinal pleura. However, on the left side, the anterior border has a distinct notch (the cardiac notch) that passes behind the 5th and 6th costal cartilages. The lower border of the lung has an excursion of as much as 5–8 cm in the extremes of respiration, but, in the neutral position (midway between inspiration and expiration), it lies along a line which crosses the 6th rib in the midclavicular line, and the 8th rib in the mid-axillary line, and which reaches the 10th rib adjacent to the vertebral column posteriorly.

The lung lobes (Figs 50, 51)

Each lung is divided by a deep *oblique fissure*, and the right lung is further divided by a *transverse fissure*. Thus, the right lung is tri-lobed, the left bi-lobed. The right oblique fissure leaves the vertebral column posteriorly at the level of the 5th rib. It follows the rough direction of the 5th rib, tending to lie slightly lower than this landmark, and ends near the costochondral junction either in the 5th space or at the level of the 6th rib. The left oblique fissure has a more variable origin, anywhere from the 3rd to 5th rib level, but its subsequent course is similar to that of the right side. When the arm is held above the head, the vertebral border of the scapula corresponds to the line of the oblique fissure – a useful and quite accurate surface marking.

The transverse fissure can be indicated by a horizontal line that runs from the 4th right costal cartilage to reach the oblique fissure in the mid-axillary line at the level of the 5th rib or interspace. These fissures are far from constant. More often than not, the transverse fissure is absent or incomplete. Often, the oblique fissure is partly fused in its upper part so that the superior segment of the lower lobe is fused with the adjacent upper lobe. Sometimes, in contrast, the individual bronchopulmonary segments (see page 63) are marked by indentations on the outer aspect of the lung that are occasionally deep enough to produce actual anomalous fissures. The upper limit of the lingula, for example, is often marked by an indentation on the anterior margin of the left upper lobe; this occasionally extends so deeply that there appears to be a left middle lobe. The apex of the right lung may be cleft by the arch of the azygos vein which, complete with its 'mesentery' of pleura, divides off a medially placed *azygos lobe* or lobule. This may be large and almost completely isolated, or may be merely a slight indentation of the upper lobe by the azygos vein.

The relationships at the root of the lung
(Figs 52, 53)

The root, or hilum, of the lung transmits the following structures within a sheath of pleura: the pulmonary artery, the two pulmonary veins, the bronchus, the bronchial vessels, lymphatics, lymph nodes and autonomic nerves. Each of these constituents is considered elsewhere, but here we must discuss their hilar relationships. These are more readily remembered logically than learned by rote.

The bronchi lie in a plane behind the heart and the roots of the great vessels – therefore the bronchus will be situated posteriorly to the pulmonary vessels. The pulmonary arteries lie along the upper borders of the atria; the pulmonary veins drain, two on each side, into the left atrium – therefore the artery must lie above the veins. The bronchial vessels hug the posterior surface of the bronchi, and this is the relationship they adopt at the hilum. Finally, the whole complex is sandwiched between the anterior and posterior pulmonary nerve plexuses.

There remains but one fact to remember: on the right side there is an additional upper lobe bronchus at the hilum which lies above ('eparterial'), but still posterior to, the pulmonary vessels.

The relationships of the lung roots themselves may be summarized thus:
1 on the left: in front, the phrenic nerve; behind, the descending aorta and the vagus nerve; above, the aortic arch; below, the pulmonary ligament;

2 on the right: in front, the superior vena cava and the phrenic nerve; behind, the vagus nerve; above, the vena azygos; below, the pulmonary ligament.

The bronchopulmonary segments
(Figs 56–58)

The concept of the anatomy of the lung has been revolutionized by the recognition that the lung is divided, functionally, not into the lobes described above, but into a series of bronchopulmonary segments each with its own bronchus, its own blood supply from the pulmonary artery and with its parenchyma distinct from adjacent segments. Lung resection surgery, postural drainage and chest radiodiagnosis are largely based on the detailed anatomy of these segments.

The arrangement of the bronchopulmonary segments varies somewhat in the two lungs but, were it not that the lingular branches arise from the upper lobe bronchus on the left, and the middle lobe branches derive from the lower part of the main bronchus on the right, the basic pattern would be essentially the same.

In the following description, the numbers in parentheses correspond with those used in Figs 56–58.

The right lung

The right main bronchus, after a course of some 2.5 cm, gives off at right angles the *upper lobe bronchus*. After a 1 cm course, this in turn trifurcates or else has a very close double bifurcation into three segmental bronchi: the *apical* (1), which passes upwards and laterally; the *posterior* (2), which passes backwards, laterally and slightly upwards; and the *anterior* (3), which passes forwards, slightly laterally and slightly downwards. The main bronchus then continues as a long (3 cm) length of primary bronchus before giving off a forward and downward directed branch that is the *middle lobe bronchus*. This is 1–1.5 cm long, and bifurcates into the lateral (4) and medial (5) divisions of the middle lobe.

The long segment of bronchus between the upper lobe bronchus and the middle lobe bronchus has not yet acquired a definitive name; it is best referred to simply as the 'lower part of the right main bronchus'. It is this portion that is crossed by the main trunk of the right pulmonary artery, hence the old term 'hyparterial bronchus' – in contrast to the right upper lobe bronchus, which is above the artery and which was termed, in older anatomical works, the 'eparterial bronchus'.

Opposite and just below the origin of the middle lobe bronchus (or occasionally on a level with it) arises the bronchus to the *apical segment of the lower lobe* (6). This bronchus passes directly backwards as a short trunk, up to 1 cm long, which then trifurcates into superior, medial and lateral branches. When a patient lies in bed, this bronchus projects directly posteriorly from the stem of the lower lobe bronchus and is therefore frequently the place in which an inhaled body or retained secretions tend to collect.

Right lateral   Left lateral

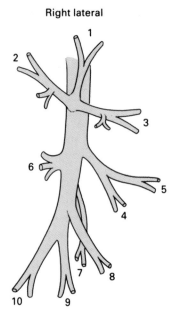

Right
Upper lobe
1 Apical bronchus
2 Posterior bronchus
3 Anterior bronchus

Middle lobe
4 Lateral bronchus
5 Medial bronchus

Lower lobe
6 Apical bronchus
7 Medial basal
  (cardiac) bronchus
8 Anterior basal
  bronchus
9 Lateral basal
  bronchus
10 Posterior basal
  bronchus

Left
Upper lobe
1 ⎫
2 ⎭ Apicoposterior bronchus
3 Anterior bronchus

Lingula
4 Superior bronchus
5 Inferior bronchus

Lower lobe
6 Apical bronchus

8 Anterior basal bronchus

9 Lateral basal bronchus

10 Posterior basal bronchus

Fig. 56 Diagram of the branches of the bronchial tree.

About 1.5 cm below the superior (apical) bronchus of the lower lobe is given off the *medial basal (or cardiac) bronchus* (7), which originates from the medial side of the main stem of the lower lobe bronchus. Then, in rapid succession, are given off the other basal bronchi:
1 the *anterior basal* (8), which runs downwards, forwards and laterally;
2 the *lateral basal* (9), which runs downwards and laterally; and
3 the *posterior basal* (10), the largest branch, which runs downwards and backwards, continuing the direction of the main lower lobe bronchus.

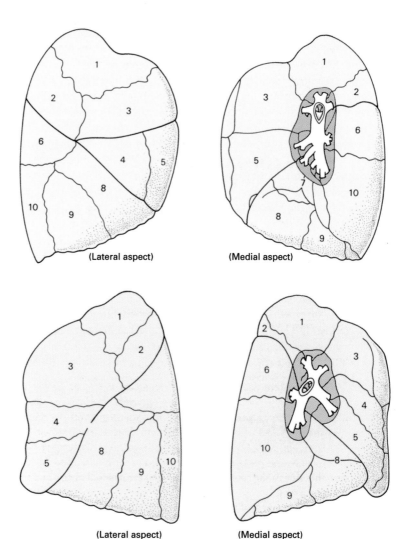

Fig. 57 The segments of the right lung. The key of the numbering is given in Fig. 56.

(Lateral aspect)   (Medial aspect)

Fig. 58 The segments of the left lung. The key of the numbering is given in Fig. 56.

(Lateral aspect)   (Medial aspect)

The left lung

The left main bronchus has a course of 5 cm before giving off the left upper lobe bronchus. The left pulmonary artery loops above this bronchus and there is thus no 'eparterial bronchus' comparable to that of the right side. The *upper lobe bronchus* passes laterally for about 1 cm, and then bifurcates into a *superior* and an *inferior* (or *lingular*) *division*. The superior division soon divides to supply the *apical, anterior* and *posterior* segments of the upper lobe just as on the right side, except that usually the apical and posterior bronchi originate by a common trunk, termed the apico-posterior bronchus (1 and 2), shortly after the separate anterior bronchus (3) is given off. Sometimes, the three bronchi arise independently, as on the right side.

The inferior division of the left upper lobe bronchus supplies the *lingula*, the tongue-like projection that constitutes the antero-inferior part of the left upper lobe. This lingular bronchus passes downwards, forwards and somewhat laterally and then bifurcates, after 1–2 cm, into *superior* (4) and *inferior* (5) branches. This division into superior and inferior segments is quite characteristic of the lingular lobe and is in contrast to the medial and lateral divisions of the middle lobe on the right side. An indentation on the anterior margin of the upper lobe frequently marks the upper limit of the lingula (Fig. 58).

The downward direction of the lingular bronchus explains the frequency with which the lingular segment is affected, together with the left lower lobe, in infections and in bronchiectasis.

The bronchi of the left lower lobe resemble the distribution on the right side except there is no medial basal (cardiac) branch.

Bronchoscopic anatomy

The segmental anatomy of the bronchial tree must now be related to the appearance at bronchoscopy (Fig. 59).

The trachea is seen as a glistening tube, which is white at the rings of cartilage and reddish between. This tube is flattened somewhat where it is crossed by the aortic arch, the pulsations of which can be observed through the tracheal wall. The tracheal bifurcation, or carina, lies a little to the left

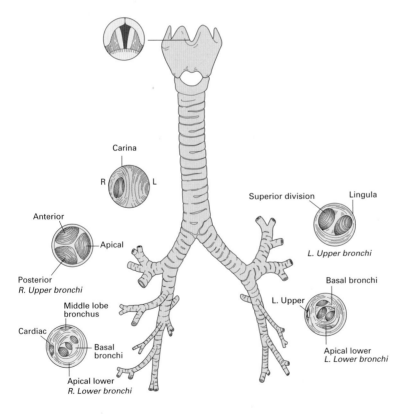

Fig. 59 The bronchoscopic anatomy of the bronchial tree.

of the mid-tracheal line, because of the more vertically situated right main bronchus; it has the appearance of a short, sharp, shining, sagittal ridge.

It is then easier to advance down the wider and more vertical right main bronchus. First, the orifice of the right upper lobe bronchus is seen on its lateral wall. Inspection of this reveals the typical division into three openings of equal size, giving a characteristic tricornate appearance. Looking into the main orifice, the orifices of its anterior, posterior and apical branches can be seen. Still advancing, a horizontal ridge appears as the anteriorly placed orifice of the middle lobe bronchus comes into view. Below this, we pass into the lower lobe bronchus. Posteriorly, its apical branch orifice can be seen, then the cardiac orifice appears on the medial side, and finally clumped close together, from above down, appear the anterior basal, lateral basal and posterior basal orifices.

Withdrawing the instrument to the trachea and now advancing it along the left main bronchus, the first feature of note is the greater length of main bronchus encountered before the orifice of the left upper lobe bronchus appears on the lateral wall. The upper lobe bronchus rapidly divides into the lingular bronchus and the left upper lobe bronchus proper. Advancing along the lower lobe bronchus brings its apical branch into view posteriorly; then, beyond this, the cluster of orifices of the anterior, lateral and posterior basal bronchi.

The structure of the lung and bronchial tree

The basic arrangement of the bronchial wall comprises mucosa, basement membrane, a submucous layer of elastic tissue, non-striated bronchial muscle and, finally, an outer fibrous coat containing cartilage.

The lining epithelium of the trachea and larger bronchi is in several layers: a basal layer, which rests on a well-defined basal membrane, an intermediate zone of spindle-shaped cells and a superficial sheet of columnar ciliated cells which are interspersed with mucus-secreting goblet cells. In chronic inflammatory conditions, the ciliated epithelium becomes replaced by stratified cells, which are non-ciliated. This metaplasia may also occur following prolonged intubation of the trachea and tracheostomy. In the finer bronchi, the epithelium becomes cuboidal and ciliated, with far fewer goblet cells. The alveoli are lined with a layer of epithelium which is so thin that, except where nuclei are present, it is often invisible in conventionally prepared histological material. Electron microscopy and special staining techniques have shown that the epithelium is, in fact, intact – although with a thickness of only 0.2 µm away from the cell nuclei – and rests on a fine basement membrane. Alveolar air is thus separated from blood in the pulmonary capillary tree by an extremely fine membrane which, nevertheless, consists of four layers: capillary wall, capillary basement membrane, alveolar basement membrane and alveolar epithelium. Among the flattened epithelial cells of the alveolar wall are others that are large and have a vacuolated appearance. These are the type II pneumocytes, which secrete surfactant – the surface-action lipoprotein complex (phospholipoprotein) that prevents the air bubbles contained in the alveoli from collapsing.

Although surfactant is present in the fetal lung at as early as 2–3 weeks, it occurs in increasing amounts until maturity. The development of surfactant can be inferred from the presence of lecithin in the amniotic fluid. Absence of surfactant because of either immaturity or genetic abnormality is responsible for the respiratory distress syndrome of the newborn.

The submucous layer of the bronchial tree consists of longitudinally disposed elastic fibres that donate the important property of elastic recoil to the air-conducting system. This layer also contains a rich capillary vascular plexus and lymphoid tissue. It is this elastic layer that produces the force of retraction, which tends to pull the lung away from the chest wall and so creates a negative intrapleural pressure. As the patient ages, the amount of elastic recoil diminishes. This results in the progressive decrease in intrapleural pressure in elderly patients and accounts for some of the decrease in pulmonary compliance and functional residual capacity.

Deep to this elastic layer is the zone of non-striated muscle. The muscle fibres of the bronchial tree form a 'geodesic network'. A geodesic line is the shortest line between two points on a surface (for example, an arc connecting two points on a sphere), and is the ideal engineering arrangement for producing or withstanding pressures within a tube, combined with least tendency for the fibres to slip along its surface. The *relative* thickness of the muscle coat increases as the branching bronchi become narrower and is in greatest proportion in the terminal bronchioles. The muscle layer forms a sphincter around the openings from the alveolar ducts that lead into the atria (see below), beyond which the muscle fibres disappear.

The cartilage rings of the extrapulmonary main bronchi are replaced in their intrapulmonary branches by irregular cartilaginous plates that are embedded in the outer fibrous coat of the bronchi. At every bronchial division, there is a saddle-shaped piece of cartilage that reinforces the two branches at their bifurcation.

The cartilages become progressively smaller and more incomplete; they finally disappear entirely in bronchioles of approximately 0.6 mm diameter.

The air spaces (Fig. 60)

The successive subdivisions of the terminal part of the bronchial tree are:
1 bronchioli;
2 respiratory bronchioli;
3 alveolar ducts;
4 atria;
5 air sacs (or alveolar sacs);
6 air cells (or alveoli).

The *bronchioles* are the finer bronchial ramifications from whose walls cartilage has disappeared; they are usually of the region of 0.6 mm in diameter. Their walls are made up largely of a comparatively thick zone of bronchial muscle and they are lined by a cuboidal ciliated epithelium with but few goblet cells. Between the epithelium and muscle is a thin elastic lamina. The *respiratory bronchioles* bear small alveoli, or air cells, on

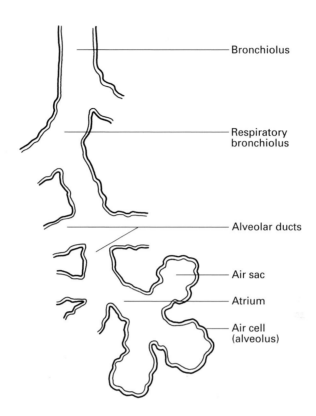

Bronchiolus

Respiratory bronchiolus

Alveolar ducts

Air sac

Atrium

Air cell (alveolus)

Fig. 60 The anatomy of the bronchial terminations.

their walls and are lined by a non-ciliated cuboidal epithelium. The distal extremity of each respiratory bronchiole is termed the *alveolar duct*. From three to six roughly spherical chambers arise from the termination of each alveolar duct. These are the *atria* which, in turn, give rise to a number of air sacs whose walls are studded with extremely thin-walled *air cells* or *alveoli*.

Each bronchiole with its subdivisions is termed a primary lung lobule.

The pulmonary blood supply

The pulmonary artery provides a capillary plexus in intimate relationship to the alveoli and is concerned solely with alveolar gas exchange. The blood supply to the lung itself, to its lymph nodes, to the bronchi and to the visceral pleura is entirely provided by the bronchial arteries. Venous drainage from the walls of the larger bronchi is carried out by the bronchial veins; the drainage of the smaller bronchi, together with that of the alveolar capillaries, is a function of the pulmonary veins. Thus, although there is no communication between the bronchial and pulmonary arteries, a good deal of blood brought to the lung by the bronchial arteries is drained by the pulmonary veins.

The *pulmonary artery* and its subdivisions closely follow the ramifications of the bronchial tree, so that each air sac has its own minute twig, which then breaks up into capillaries that form the richest vascular network in the body. The *pulmonary vein* tributaries are derived partly from the capillaries of the pulmonary artery and partly from those of the bronchial artery. Unlike the branches of the pulmonary artery, which run in close relation to the bronchial tree, the venous tributaries lie between the lung segments, thus providing a valuable landmark to the surgeon in the performance of segmental resections. At the apex of each bronchopulmonary segment, the pulmonary vein draining that segment meets the segmental artery and passes alongside it to the hilum.

There are two main pulmonary veins on each side, which drain separately into the left atrium. On the left side, the upper and lower lobe each have their own main pulmonary vein; on the right side, the upper and middle lobes share the upper pulmonary vein, the lower lobe drains via the lower vein.

The *bronchial arteries* are to the lungs what the hepatic artery is to the liver: they supply the actual pulmonary stroma – the bronchi, lung tissue, visceral pleura and pulmonary nodes.

The arteries are variable in both number and origin; there are usually three, one for the right lung and two for the left. The left bronchial arteries usually arise from the anterior aspect of the descending thoracic aorta. The right artery is more variable; it may arise from the aorta, the 1st intercostal artery, the 3rd intercostal artery (which is the 1st intercostal branch of the aorta), the internal thoracic artery or the right subclavian artery. Occasionally, all three arteries arise from a common trunk derived from the aorta.

The arteries lie against the posterior walls of their respective bronchi. They follow and supply the bifurcating bronchial tree as far as the small bronchioles but disappear as soon as alveoli appear in the walls of the ducts; all available respiratory epithelium is thus supplied from the pulmonary arterial tree.

The *bronchial veins* are usually two in number on each side; the right drain into the azygos vein, the left into the superior hemiazygos or the left superior intercostal vein. They only drain blood from the first two or three bifurcations of the bronchial tree; more distally the bronchial arterial blood drains into the radicles of the pulmonary veins.

The bronchial blood flow, together with that in the venae cordis minimae (Thebesian veins) of the heart, constitutes a physiological shunt whereby venous blood is mixed with arterial blood in the heart. An increase in bronchial blood flow may occur during acute pulmonary infections and bronchiectasis and will inevitably increase the shunt effect. Normally, this shunt of venous blood to the left side of the heart constitutes less than 1–2% of the cardiac output; this is the so-called 'physiological shunt'. In the normal individual, this shunt is increased by minimal ventilation/perfusion mismatching in the lung and may then total 5% of the cardiac output.

The fine arrangement of the blood vessels within the bronchial wall is of some practical interest. The arterial plexus derived from the bronchial artery lies external to the bronchial muscle; vessels pierce the muscle coat to form a capillary plexus in the submucosa. The venous radicles, in turn, pierce the muscle layer in order to drain into the venous plexus in the areolar tissue outside the muscle. Blood must therefore traverse the bronchial muscle both to reach and to leave the submucous capillary plexus. Oedema of the bronchial wall will occlude the low-pressure veins before the high-pressure arteries; the resultant venous obstruction produces further mucosal swelling and thus accentuates the bronchial obstruction.

Lymphatics

A superficial lymphatic plexus drains the visceral pleura; a deep plexus, lying alongside the pulmonary vessels, drains the bronchi but does not reach beyond the alveolar ducts into the more distal air spaces. Both lymphatic plexuses drain into bronchopulmonary lymph nodes placed at the points of bifurcation of the larger bronchi. Thence, lymph drains to the tracheobronchial nodes, which, in turn, empty into the right and left bronchomediastinal trunks. The right trunk may drain into the right lymph duct and the left may empty into the thoracic duct. More often, they open directly and independently into the junction between the internal jugular and the subclavian veins on either side.

Innervation

Sympathetic (T2–4) and parasympathetic (vagal) fibres form a posterior pulmonary plexus at the root of the lung. Fibres pass thence around the lung root to form an anterior pulmonary nerve plexus. Fibres stream from these plexuses into the lung along the blood vessels and bronchi.

The mucous glands are supplied by secretomotor parasympathetic fibres. The bronchial muscles receive bronchodilator (inhibitory) fibres from the sympathetic system and bronchoconstrictor fibres from the vagus. The bronchial vessels are under sympathetic vasomotor control, which is much less in evidence in the case of the thin-walled pulmonary vascular bed; this is little affected by sympathetic stimulation and is mainly passively controlled, e.g. by right ventricular pressure.

Afferent fibres, sensitive to stretch, are transmitted from the lung via the vagus to the medullary respiratory centre.

The development of the respiratory tract

In the early fetus (the 3 mm embryo), a median ventral diverticulum appears in the foregut, which is termed the tracheobronchial groove. This gradually deepens and its edges nip together caudally so that the diverticulum becomes separated from the primitive oesophagus except at the laryngeal aditus. Meantime, the caudal prolongation of the diverticulum divides into the two main bronchi and further proliferation results in the formation of the lung bud on each side. A persistent tracheo-oesophageal fistula may occur as an embryonic anomaly, which indicates the close developmental relationship between the foregut and the respiratory

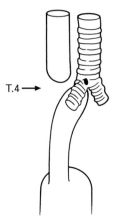

T.4 →

Fig. 61 The usual anatomy of a congenital tracheo-oesophageal fistula: the upper oesophagus ends blindly; the lower oesophagus is connected with the trachea at the level of the 4th thoracic vertebra.

passages; it is usually associated with atresia of the oesophagus, the fistula being situated below the atretic segment (Fig. 61).

The diaphragm

The diaphragm constitutes the great muscular septum between the thorax and the abdomen; it is one of the distinguishing features of mammalian anatomy.

Anatomical features (Fig. 62)

The diaphragm consists of peripheral muscle with a central trefoil-shaped tendon of strong interlacing bundles that blend above with the fibrous pericardium. The muscle takes a complex origin from the crura, the arcuate ligaments, the costal margin and the xiphoid.

The *crura* arise from the lumbar vertebral bodies; the left from the 1st and 2nd, the larger right from the 1st, 2nd and 3rd.

The *arcuate ligaments* are the *median*, which is a fibrous arch joining the two crura, the *medial*, which is a thickening of the fascia over the psoas, and the *lateral*, which is a condensation of fascia over quadratus lumborum ending laterally near the tip of the 12th rib.

The *costal origin* is from the tips of the last six costal cartilages.

The *xiphoid origin* comprises two slips from the posterior aspect of the xiphoid.

The diaphragmatic foramina (Figs 62, 63)

The three major openings are for:
1 the inferior vena cava, at the level of the body of the 8th thoracic vertebra;

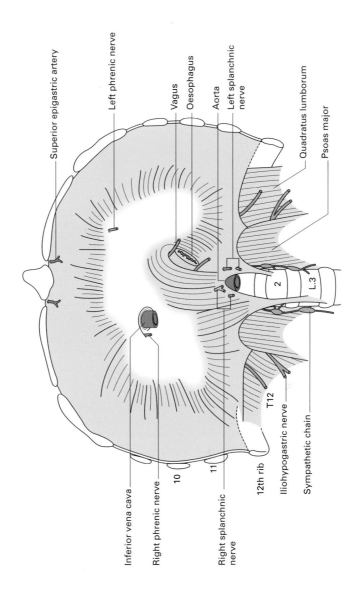

Fig. 62 The abdominal aspect of the diaphragm.

Superior epigastric artery

Left phrenic nerve

Vagus

Oesophagus

Aorta

Left splanchnic nerve

Quadratus lumborum

Psoas major

Inferior vena cava

Right phrenic nerve

10

11

Right splanchnic nerve

12th rib

Iliohypogastric nerve

Sympathetic chain

T12

2

L.3

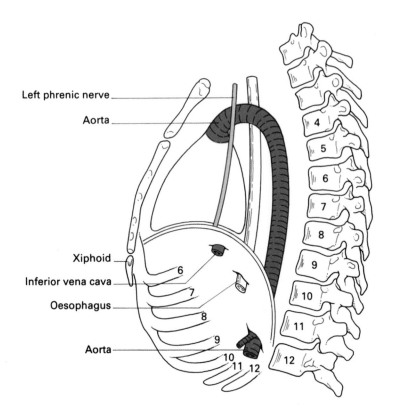

Left phrenic nerve

Aorta

Xiphoid

Inferior vena cava

Oesophagus

Aorta

Fig. 63 Lateral view of the diaphragm to show the levels at which it is pierced by major structures.

2 the oesophagus, together with the vagi and the oesophageal branches of the left gastric vessels, at T10;

3 the aorta, together with the thoracic duct and azygos vein, behind the median arcuate ligament at T12.

In addition, the sympathetic trunk passes behind the medial arcuate ligament, the splanchnic nerves pierce the crura, the hemiazygos vein drains through the left crus, the superior epigastric vessels pass between the xiphoid and costal origins of the diaphragm into the posterior rectus sheath, the lower intercostal nerves and vessels enter the anterior abdominal wall between the interdigitations of the diaphragm and transversus abdominis and lymphatics stream from the retroperitoneal tissues through the diaphragm to the mediastinum.

The oesophageal hiatus is reinforced by a sling of muscle fibres from the right crus, which probably plays a part in maintaining competence at the oesophagogastric junction. Not uncommonly, the left crus supplies some of the fibres forming this sling, which is occasionally derived solely from it.

Nerve supply

The phrenic nerve (C3–5; see page 164 for a detailed description) provides the motor supply of the diaphragm, apart from an unimportant contribution to the crura from T11 and T12. Section of the phrenic nerve is followed by complete atrophy of the corresponding hemidiaphragm. The phrenic nerve also transmits proprioceptive fibres from the centre of the diaphragm, although the periphery of this muscle has its sensory supply from the lower thoracic nerves.

The right phrenic nerve pierces the central tendon to the lateral side of the inferior vena cava (some fibres may actually accompany the vein through its foramen). The left nerve pierces the muscle about 1 cm lateral to the attachment of the pericardium. The terminal fibres of each nerve supply the muscle on its abdominal aspect.

The diaphragm as a muscle of respiration

The apex of the dome of the diaphragm reaches the level of the 5th rib in the mid-clavicular line, i.e. it is level with a point about 2.5 cm below the nipple. The right hemidiaphragm is rather higher than the left, and both domes rise somewhat in the horizontal position. When the subject lies on his/her side, the upper cupola sinks to a lower level than its partner and its movements are relatively diminished. The level of the diaphragm is elevated in late pregnancy, gross ascites or obesity, in pneumoperitoneum and in patients with large abdominal tumours; such subjects all have some degree of respiratory limitation.

In inspiration, the diaphragm moves vertically downwards (the domes considerably more than the central tendon), and this has a piston-like action in enlarging the thoracic cavity. A subsidiary effect is that the lower costal margin is raised and everted with consequent expansion of the base of the thorax.

In expiration, the diaphragm relaxes; in forced expiration, it is actually pushed upwards by the increased intra-abdominal pressure effected by contraction of the muscles of the anterior abdominal wall.

It is estimated that the movement of the diaphragm accounts for some 60–75% of the total tidal volume of respiration; in some subjects during quiet breathing, it may, indeed, be the only functioning muscle in inspiration. It is therefore interesting that bilateral phrenic interruption with complete diaphragmatic paralysis may cause little respiratory difficulty, providing the lungs are relatively normal. In quiet respiration, the diaphragm has a range of movement of 1.5 cm. In deep breathing this increases to 7–13 cm.

In addition to its important role as a muscle of respiration, the diaphragm helps increase the intra-abdominal pressure in defecation, micturition, vomiting and parturition, as well as taking part in the mechanism of the 'cardiac sphincter' (see below).

The diaphragm and the 'cardiac sphincter'

At the cardio-oesophageal junction, there exists a rather extraordinary sphincter mechanism that allows food and liquids to pass readily into the stomach, which prevents free regurgitation into the oesophagus even when standing on one's head or in forced inspiration (when there is a pressure difference of approximately 80 mmHg between the intragastric and intra-oesophageal pressures), but which can relax readily to allow vomiting or belching to occur.

In spite of extensive investigations, the exact nature of this sphincter is not understood. It is probably a complex affair made up of:

1 a physiological muscular sphincter at the lower end of the oesophagus;
2 a plug-like action of the mucosal folds at the cardia;
3 a valve-like effect of the obliquity of the oesophagogastric angle;
4 a diaphragmatic sling that maintains the normal position of the cardia and has a pinch-cock action on the lower oesophagus;
5 the positive intra-abdominal pressure that tends to squeeze the walls of the intra-abdominal portion of the oesophagus together.

Although a true anatomical sphincter cannot be shown by dissection, a physiological sphincter can be deduced from the high-pressure zone demonstrated within the lower oesophagus, which disappears when the oesophageal muscle is divided, as in Heller's operation for achalasia of the cardia. Reinforcing the sphincter are the mucosal folds of the cardia, which act as a plug wedged within the muscular ring.

The crural sling of the diaphragm around the lower oesophagus is important in maintaining the normal position of the cardio-oesophageal junction below the diaphragm. If the hiatus is enlarged and lax, the stomach can slide upwards into the chest (a 'sliding hiatus hernia') and the normal valve-like angle between oesophagus and cardia straightens out.

There also appears to be a definite pinch-cock mechanism on the oesophagus when the diaphragm contracts in full inspiration – a phase at which intrathoracic pressure is lowest, intra-abdominal pressure highest and conditions most favourable for fluids to be forced at high pressure upwards through the cardiac orifice. The diaphragm is an important but not essential part of the cardiac sphincter mechanism, since a sliding hiatus hernia is not necessarily accompanied by regurgitation providing the physiological sphincter is competent. Similarly, free regurgitation occurs in some subjects with an apparently normal oesophageal hiatus, presumably because of some defect in the function of the physiological sphincter.

The development of the diaphragm

The diaphragm is formed (Fig. 64) by fusion in the fetus of:

1 the septum transversum, which constitutes the central tendon;
2 the dorsal oesophageal mesentery;
3 a peripheral rim derived from the body wall;
4 the pleuroperitoneal membranes, which close the primitive communications between the pleural and peritoneal cavities.

The septum transversum is the mesoderm which, in early development, lies in front of the head end of the embryo. With the folding off of the head, this mesodermal mass is carried ventrally and caudally, to lie in its definitive position at the anterior part of the diaphragm. During this migration, the cervical myotomes and nerves contribute muscle and nerve supply,

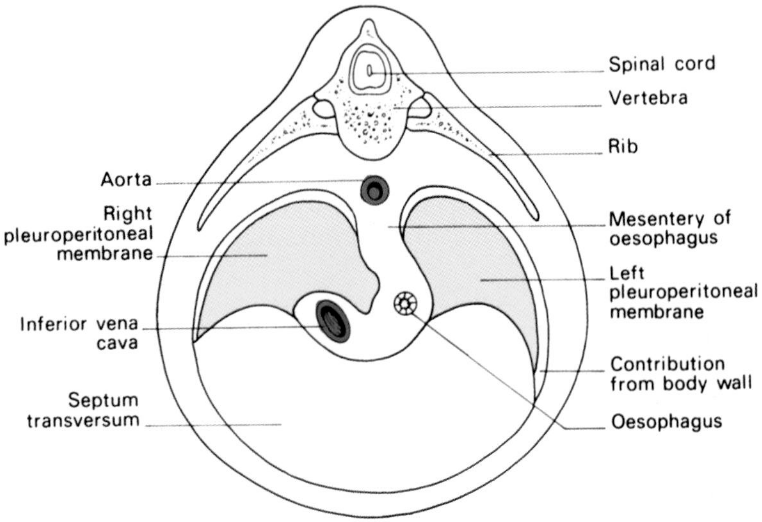

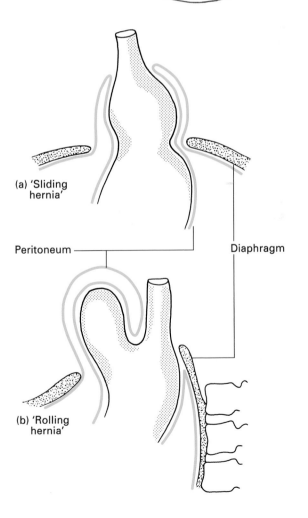

Fig. 64 The development of the diaphragm.

Spinal cord
Vertebra
Rib
Aorta
Right pleuroperitoneal membrane
Mesentery of oesophagus
Left pleuroperitoneal membrane
Inferior vena cava
Contribution from body wall
Septum transversum
Oesophagus

(a) 'Sliding hernia'

Peritoneum

Diaphragm

(b) 'Rolling hernia'

Fig. 65 (a) A sliding hiatus hernia and (b) a rolling hiatus hernia.

respectively, thus accounting for the long course of the phrenic nerve from the neck to the diaphragm.

In spite of such a complex story, congenital abnormalities of the diaphragm are unusual. However, a number of defects may occur, giving rise to a variety of congenital herniae through the diaphragm. These may be:

1 through the foramen of Morgagni – anteriorly between the xiphoid and costal origins;
2 through the foramen of Bochdalek – the pleuroperitoneal canal – lying posteriorly;
3 through a deficiency of the whole central tendon (occasionally such a hernia may be traumatic in origin);
4 through a congenitally large oesophageal hiatus.

Far more common are the acquired hiatus herniae, divided into sliding hernia and rolling hernia. These occur in patients usually of middle age, in whom weakening and widening of the oesophageal hiatus has occurred (Fig. 65).

In the *sliding hernia*, the upper stomach and lower oesophagus slide upwards into the chest through the lax hiatus when the patient lies down or bends over. In the *rolling hernia* (which is far less common), the cardia remains in its normal position and the cardio-oesophageal junction is intact, but the fundus of the stomach rolls up through the hiatus in front of the oesophagus, hence the alternative term of para-oesophageal hernia.

Part 2
The Heart and Great Veins of the Neck

The pericardium

The heart and the roots of the great vessels are enclosed in the conical fibrous pericardium. The apex of this sac is fused with the adventitia of the great vessels at the level of the angle of Louis (the manubriosternal junction). The anterior surface is attached by loose fibrous tissue, termed the *sternopericardial ligament*, to the posterior aspect of the sternum. Inferiorly, the base blends with the central tendon of the diaphragm. Because of this, the heart's position depends on that of the diaphragm; the heart is dragged downwards on deep inspiration, and a relatively transverse lie of the heart is associated with the high-placed diaphragm of pregnancy, of abdominal distension and of the subject with a short stocky build. For the same reason, the heart is high in recumbency and low in the erect position.

The pericardium is related anteriorly to the body of the sternum, to the 3rd–6th costal cartilages on either side, and to the thin anterior borders of both lungs. Laterally lie the mediastinal pleura and the phrenic nerve, while posteriorly are the oesophagus, descending aorta, the bronchi and the bodies of the 5th–8th thoracic vertebrae.

Within the fibrous pericardial sac lies the serous pericardium. This is no exception to the other serous membranes – the pleura, peritoneum and tunica vaginalis – in that it is produced by the invagination of a viscus (i.e. the heart) into a fetal serous sac, with the consequent formation of a double membrane. The visceral layer, or epicardium, closely adheres to the heart, the parietal layer lines the fibrous pericardium, while between the visceral and parietal layer is the pericardial cavity which, in health, has no contents.

The parietal layer is reflected around the roots of the great vessels to become continuous with the visceral layer. These lines of reflection are marked on the posterior aspect of the heart by (Fig. 66):

1 the *transverse sinus*, which lies between the superior vena cava and left atrium posteriorly and the pulmonary trunk and aorta in front;

2 the *oblique sinus*, which is bordered by the two right and two left pulmonary veins, reinforced below and on the right by the inferior vena cava, and which forms a recess between the left atrium and the pericardium.

The composition and relationships of these sinuses are readily explained embryologically. The transverse sinus is formed as a result of the S-shaped kinking of the originally tubular fetal heart (Fig. 67), so that the ventricle and truncus (which later splits into the aorta and pulmonary trunk) come to lie in front of the atrium and cava, with the transverse sinus between. The fetal dorsal mesocardium transmits a single pulmonary vein that receives tributaries from both lungs and drains into the left atrium. This pulmonary vein stem is progressively absorbed into the atrium so that the pericardial space eventually transmits two right and two left veins, leaving the oblique sinus as a recess between the pericardium posteriorly and the left atrium anteriorly.

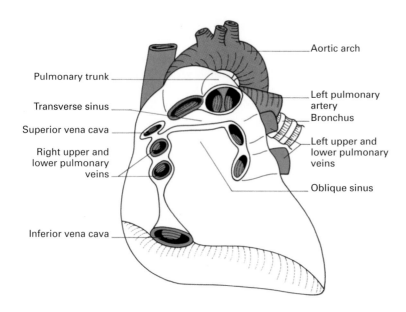

Aortic arch

Pulmonary trunk

Transverse sinus

Superior vena cava

Right upper and lower pulmonary veins

Left pulmonary artery

Bronchus

Left upper and lower pulmonary veins

Oblique sinus

Inferior vena cava

Fig. 66 The transverse and oblique sinuses of the pericardium. The heart has been removed from the pericardial sac, which is seen in anterior view.

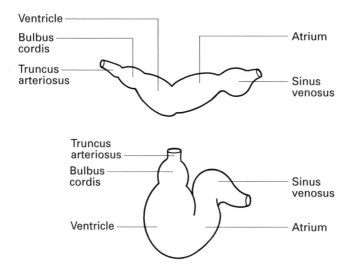

Ventricle

Bulbus cordis

Truncus arteriosus

Atrium

Sinus venosus

Truncus arteriosus

Bulbus cordis

Ventricle

Sinus venosus

Atrium

Fig. 67 Coiling of the primitive heart tube into its definitive form.

The heart

The heart is irregularly conical in shape, and is placed obliquely in the middle of the mediastinum. The right border is formed entirely by the right atrium, the left border partly by the auricular appendage of the left atrium but mainly by the left ventricle, and the inferior border chiefly by the right ventricle but also by the lower part of the right atrium and the apex of the left ventricle.

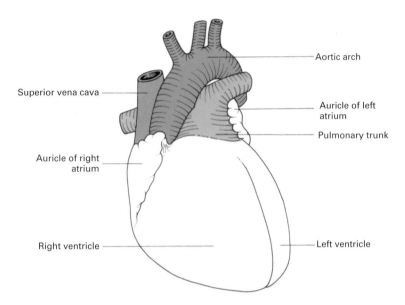

Aortic arch

Superior vena cava

Auricle of left
atrium

Pulmonary trunk

Auricle of right
atrium

Right ventricle

Left ventricle

Fig. 68 The heart in
anterior view.

The bulk of the anterior surface (Fig. 68) is formed by the right ventricle,
which is separated from the right atrium by the vertical atrioventricular
groove, and from the left ventricle by the anterior interventricular groove.
The inferior or diaphragmatic surface consists of the right and left ventricles separated by the posterior interventricular groove and the portion of
the right atrium that receives the inferior vena cava. The base, or posterior
surface (Fig. 69), is quadrilateral in shape and is formed mainly by the left

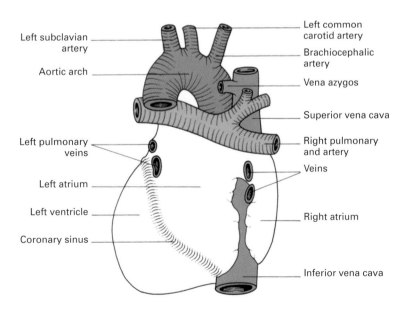

Left common
carotid artery

Left subclavian
artery

Brachiocephalic
artery

Aortic arch

Vena azygos

Superior vena cava

Left pulmonary
veins

Right pulmonary
and artery

Veins

Left atrium

Left ventricle

Right atrium

Coronary sinus

Inferior vena cava

Fig. 69 The heart in
posterior view.

atrium with the openings of the pulmonary veins and, to a lesser extent, by the right atrium.

The chambers of the heart

The right atrium (Fig. 70a) receives the superior vena cava in its upper and posterior part, the inferior vena cava and coronary sinus in its lower

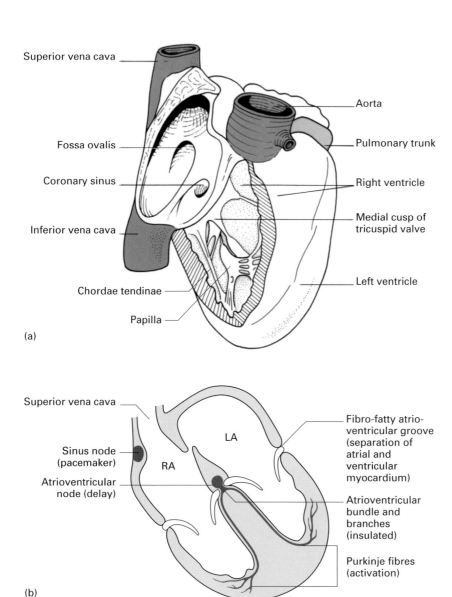

Fig. 70 (a) The interior of the right atrium and ventricle. (b) The conducting system of the heart. LA, left atrium; RA, right atrium.

part, and the anterior cardiac vein (draining much of the front of the heart) anteriorly. Running more or less vertically downwards between the venae cavae is a distinct ridge, the *crista terminalis* (indicated on the outer surface of the atrium by a shallow groove – the *sulcus terminalis*). This ridge separates the smooth-walled posterior part of the atrium, derived from the sinus venosus, from the rough-walled anterior portion, which is prolonged into the *auricular appendage* and which is derived from the fetal atrium.

The openings of the inferior vena cava and the coronary sinus are guarded by rudimentary valves; that of the inferior vena cava being continuous with the annulus ovalis around the shallow depression on the atrial septum, the *fossa ovalis*, which marks the site of the fetal foramen ovale.

The right ventricle (Fig. 70a) communicates with the right atrium by way of the vertically disposed *tricuspid valve*, and with the pulmonary trunk through the *pulmonary valve*. The tricuspid valve admits three fingers and bears three flap-like cusps (medial, anterior and inferior) that are triangular in shape and attached by their base to the fibrous ring of the tricuspid orifice. The pulmonary valve also bears three cusps, named the posterior, right anterior and left anterior, respectively.

A muscular ridge, the infundibuloventricular crest, lying between the atrioventricular and pulmonary orifices, separates the 'inflow' and 'outflow' tracts of the ventricle. The inner aspect of the inflow tract path is marked by the presence of a number of irregular muscular elevations (*trabeculae carneae*) from some of which the *papillary muscles* project into the lumen of the ventricle and find attachment to the free borders of the cusps of the tricuspid valve by way of the *chordae tendineae*. The *moderator band* is a muscular bundle crossing the ventricular cavity from the interventricular septum to the anterior wall and is of some importance since it conveys the right branch of the atrioventricular bundle to the ventricular muscle.

The outflow tract of the ventricle, the *infundibulum*, is smooth-walled and is directed upwards and to the right towards the pulmonary trunk.

The left atrium is rather smaller than the right but has somewhat thicker walls. On the upper part of its posterior wall it presents the openings of the four pulmonary veins, and on its septal surface there is a shallow depression corresponding to the fossa ovalis of the right atrium. As on the right side, the main part of the cavity is smooth-walled but the surface of the auricle is marked by a number of ridges caused by the underlying pectinate muscles.

The left ventricle (Fig. 71) communicates with the left atrium by way of the *mitral valve*, which is large enough to admit two fingers; it possesses a large anterior and a smaller posterior cusp attached to papillary muscles by chordae tendineae. With the exception of the fibrous vestibule immediately below the *aortic orifice*, the wall of the left ventricle is marked by thick trabeculae carneae.

The aortic orifice is guarded by the three semilunar cusps of the aortic valve (right posterior, left posterior and anterior), immediately above

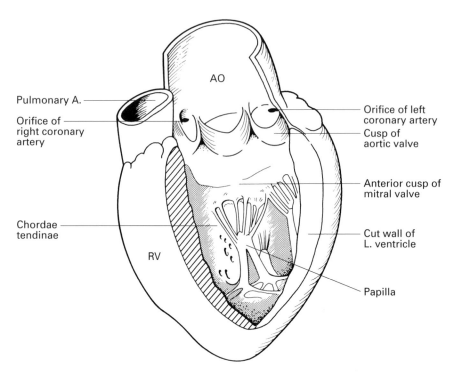

Fig. 71 The interior of the left ventricle. AO, aorta; RV, right ventricle.

which are the dilated aortic sinuses. The mouths of the right and left coronary arteries are seen in the anterior and left posterior sinus, respectively.

CLINICAL NOTE

Transoesophageal echocardiography and pulmonary artery catheters

Transoesophageal echocardiography is increasingly used in the operating room and in the intensive care unit. It gives information on all cardiac structures and their functional status, providing real-time information on morphology dimensions and wall motion to the intensivist/the anaesthetist. It is also highly sensitive in detecting thrombi in the left atrium and can give information on the functional status of the aorta and pulmonary arteries. It is often used in high-risk patients undergoing cardiac surgery, such as coronary artery bypass grafting and valve replacement, and also in thoracoabdominal aneurysm surgery (Fig. 72a,b).

Pulmonary artery catheters are used less now than in the past but still have a role in the intensive care unit and in selected patients in the operating theatre. The catheter is advanced through the superior vena cava to the right heart and thence to the pulmonary artery (Fig. 73).

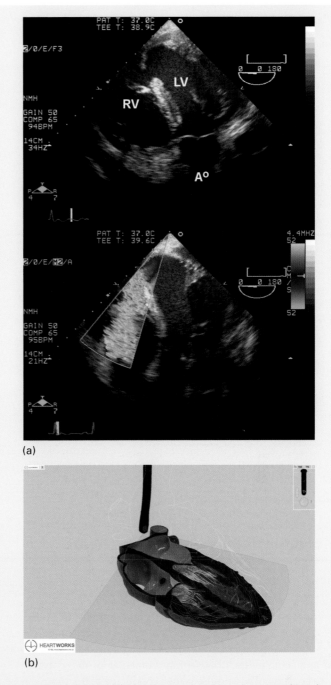

(a)

(b)

Fig. 72 (a) Colour image of the heart as produced by a transoesophageal electrocardiogram. Ao, aorta; LV, left ventricle; RV, right ventricle. (b) A three-dimensional representation of the heart with a transoesophageal echocardiogram probe *in situ*. © Heartworks. Reproduced with permission from Inventive Medical Ltd.

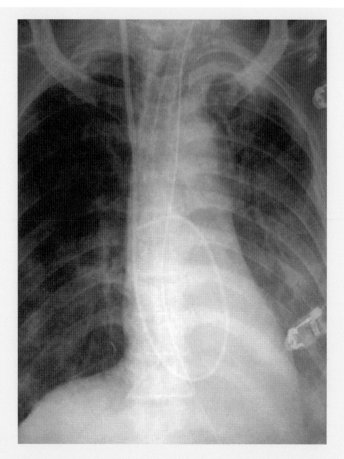

Fig. 73 An anteroposterior chest radiograph showing a pulmonary artery catheter *in situ*.

The conducting system of the heart

This consists of specialized cardiac muscle found in the sinoatrial node and in the atrioventricular node and bundle (Fig. 70b). The heart beat is initiated in the *sinoatrial node* (the 'pacemaker of the heart'), situated in the upper part of the crista terminalis just to the right of the opening of the superior vena cava into the right atrium. From there the cardiac impulse spreads throughout the atrial musculature to reach the *atrioventricular node* lying in the atrial septum immediately above the opening of the coronary sinus. The impulse is then conducted to the ventricles by way of the specialized tissue of the *atrioventricular bundle* (of His). This bundle divides at the junction of the membranous and muscular parts of the interventricular septum into its right and left branches, which run immediately beneath the endocardium to activate all parts of the ventricular musculature.

The *sinoatrial node* is supplied by the right coronary artery in approximately 65% of cases. In the remainder, the supply is usually from the circumflex branch of the left coronary artery.

The *atrioventricular node* is perfused by the right coronary artery in 80% of subjects.

The blood supply of the heart

The arterial supply to the cardiac musculature is derived from the right and left coronary arteries (Fig. 74).

The *right coronary artery* arises from the anterior aortic sinus and passes forwards between the pulmonary trunk and the right atrium to descend in the right part of the atrioventricular groove. At the inferior border of the heart it continues along the atrioventricular groove to anastomose with the left coronary at the inferior interventricular groove. It gives off a *marginal branch* along the lower border of the heart and an *interventricular branch* that runs forwards in the inferior interventricular groove to anastomose near the apex of the heart with the corresponding branch of the left coronary artery.

The left *coronary artery*, which is larger than the right, arises from the left posterior aortic sinus. Passing first behind and then to the left of the pulmonary trunk, it reaches the left part of the atrioventricular groove in which it runs laterally round the left border of the heart (the *circumflex artery*) to reach the inferior interventricular groove. Its most important branch is the anterior interventricular artery, which supplies the anterior aspect of both ventricles and passes around the apex of the heart to anastomose with the interventricular branch of the right artery.

The bulk of the venous drainage of the heart (Fig. 75) is by veins that accompany the coronary arteries and that open into the right atrium. The rest of the blood drains by means of small veins (*venae cordis minimae*) directly into the cardiac cavity.

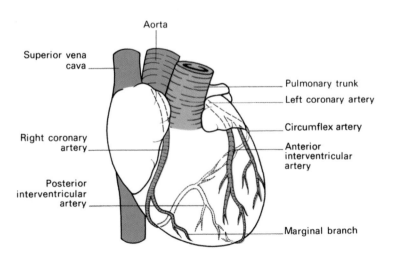

Fig. 74 The coronary arteries.

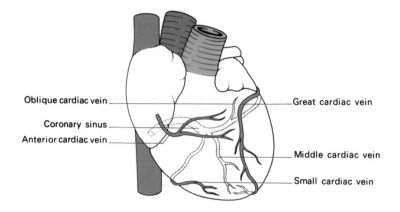

Oblique cardiac vein

Coronary sinus

Anterior cardiac vein

Great cardiac vein

Middle cardiac vein

Small cardiac vein

Fig. 75 The venous drainage of the heart.

The *coronary sinus* lies in the posterior atrioventricular groove and opens into the right atrium just to the left of the mouth of the inferior vena cava. It receives the following:

1 the *great cardiac vein* in the anterior interventricular groove;
2 the *middle cardiac vein* in the inferior interventricular groove;
3 the *small cardiac vein* that accompanies the marginal artery along the lower border of the heart;
4 the *oblique vein* that descends obliquely on the back of the left atrium and that opens near the left extremity of the coronary sinus.

The *anterior cardiac vein* crosses the anterior atrioventricular groove and its contained right coronary artery; it drains much of the anterior surface of the heart and opens directly into the right atrium.

Nerve supply

The nerve supply of the heart is derived from the vagus (cardio-inhibitor) and the cervical and upper thoracic sympathetic ganglia (cardio-accelerator) by way of the superficial and deep cardiac plexuses (see page 236).

Surface markings

The outline of the heart can be represented on the surface by an irregular quadrangle bounded by the following four points (Fig. 76):

1 the 2nd left costal cartilage 1.25 cm from the edge of the sternum;
2 the 3rd right costal cartilage 1.25 cm from the sternal edge;
3 the 6th right costal cartilage 1.25 cm from the sternum;
4 the 5th left intercostal space 9 cm from the midline (corresponding to the apex beat).

The left border of the heart is formed almost entirely by the left ventricle; the lower border corresponds to the right ventricle and the apical part of the left ventricle; and the right border is formed by the right atrium.

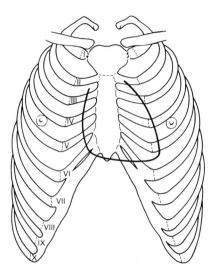

Fig. 76 The surface markings of the heart.

CLINICAL NOTE

The radiographic anatomy of the heart and great vessels: posteroanterior radiographs (Fig. 77)

The greater part of the mediastinal shadow in a posteroanterior film of the chest is formed by the heart and the great vessels. Normally, the transverse diameter of the cardiac shadow should not exceed half the total width of

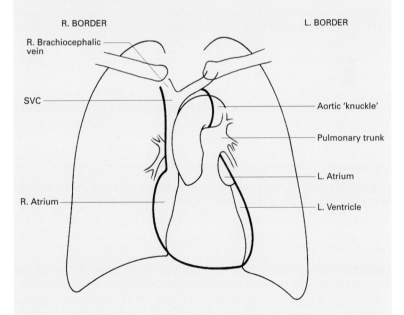

Fig. 77 Tracing of a posteroanterior radiograph of the chest. The right and left borders of the mediastinal shadow have been emphasized. SVC, superior vena cava.

the chest, but since it varies widely with body build and the position of the heart, these factors must also be assessed. The shape of the cardiac shadow also varies a good deal with the position of the heart, being long and narrow in a vertically disposed heart and broad and rounded in the so-called horizontal heart.

The right border of the mediastinal shadow is formed from above downwards by the right brachiocephalic vein, the superior vena cava and the right atrium. Immediately above the heart, the left border of the mediastinal shadow presents a well-marked projection, the aortic knuckle, which represents the arch of the aorta seen 'end-on'. Beneath this there are successively the shadows due to the pulmonary trunk (or the infundibulum of the right ventricle), the auricle of the left atrium and the left ventricle. The shadow of the inferior border of the heart blends centrally with that of the diaphragm, but on either side the two shadows are separated by the well-defined cardiophrenic angles.

Developmental anatomy
The development of the heart

The primitive heart is a single tube that soon shows grooves demarcating the sinus venosus, atrium, ventricle and bulbus cordis from behind forward. As this tube enlarges, it kinks so that its caudal end, receiving venous blood, comes to lie behind its cephalic end with its emerging arteries (Fig. 67). The sinus venosus is later absorbed into the atrium and the bulbus becomes incorporated into the ventricle so that, in the fully developed heart, the atria and great veins come to lie posterior to the ventricles and the roots of the great arteries.

The boundary tissue between the primitive single atrial cavity and single ventricle grows out as a *dorsal* and a *ventral endocardial cushion*, which meet in the midline, thus dividing the common atrioventricular orifice into a right (tricuspid) and left (mitral) orifice.

The division of the primitive atrium into two is a complicated process but an important one in the understanding of congenital septal defects (Fig. 78). A partition, the *septum primum*, grows downwards from the posterior and superior walls of the primitive common atrium to fuse with the endocardial cushions. Before fusion is complete, a hole appears in the upper part of this septum that is termed the *foramen secundum* in the septum primum.

A second membrane, the *septum secundum*, then develops to the right of the primum but this is never complete; it has a free lower edge that does, however, extend low enough for this new septum to overlap the foramen secundum in the septum primum and hence to close it.

These two overlapping defects in the septa form the valve-like *foramen ovale*, which shunts blood from the right to the left heart in the fetus (see below). After birth, this foramen usually becomes completely fused,

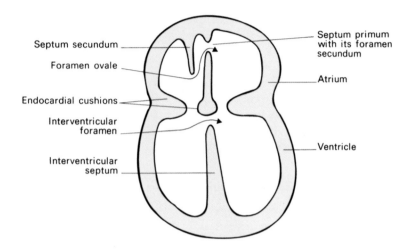

Fig. 78 The development of the chambers of the heart. Note the septum primum and septum secundum, which form the interatrial septum and which leave the foramen ovale as a valvular opening between them.

leaving only the fossa ovalis on the septal wall of the right atrium as its memorial. In approximately 10% of adult subjects, however, a probe can still be insinuated through an anatomically patent, although functionally sealed, foramen.

Division of the ventricle is commenced by the upgrowth of a fleshy septum from the apex of the heart towards the endocardial cushions. This stops short of dividing the ventricle completely and thus it has an upper free border, forming a temporary interventricular foramen. At the same time, the single truncus arteriosus is divided into the aorta and pulmonary trunk by a spiral septum (hence the spiral relations of these two vessels), which grows downwards to the ventricle and fuses accurately with the upper free border of the ventricular septum. This contributes the small *pars membranacea septi*, which completes the separation of the ventricle in such a way that blood on the left of the septum flows into the aorta and blood on the right into the pulmonary trunk.

The primitive sinus venosus becomes absorbed into the right atrium so that the venae cavae draining into the sinus come to open separately into this atrium. The smooth-walled part of the adult atrium represents the contribution of the sinus venosus; the pectinate trabeculated part represents the portion derived from the primitive atrium. Rather similarly, the adult left atrium has a double origin. The original single pulmonary venous trunk entering the left atrium becomes absorbed into it, and donates the smooth-walled part of this chamber with the pulmonary veins, entering as four separate openings; the trabeculated part of the definitive left atrium is the remains of the original atrial wall.

The development of the aortic arches and their derivatives (Fig. 79)

Emerging from the bulbus cordis is a common arterial trunk, termed the *truncus arteriosus*, from which arise six pairs of aortic arches, equivalent to

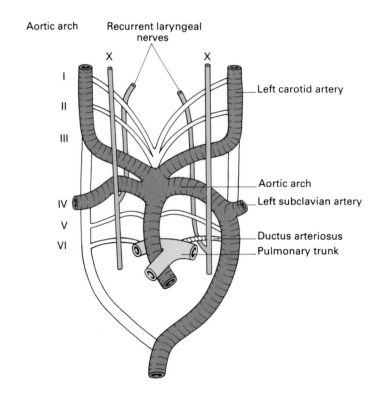

Aortic arch   Recurrent laryngeal
                            nerves

I

II

III

Left carotid artery

IV

Aortic arch

Left subclavian artery

V

VI

Ductus arteriosus

Pulmonary trunk

Fig. 79 The primitive aortic arches and their adult derivatives. The parts of the arches that disappear are shown in white.

the arteries supplying the gill clefts of fish. These arteries curve dorsally around the pharynx on either side and join to form two longitudinally placed dorsal aortae that fuse distally into the descending aorta.

The 1st and 2nd arches disappear; the 3rd arches become the carotid arteries. The 4th arch on the right becomes the brachiocephalic artery and right subclavian artery; on the left, it differentiates into the definitive aortic arch, gives off the left subclavian artery and links up distally with the descending aorta. The 5th arch artery is rudimentary and disappears. When the truncus arteriosus splits longitudinally to form the ascending aorta and pulmonary trunk, the 6th arch (unlike the others) remains linked with the latter and forms the right and left pulmonary arteries. On the left side, this arch retains its connection with the dorsal aorta to form the ductus arteriosus (the ligamentum arteriosum of adult anatomy).

This asymmetrical development of the aortic arches accounts for the different course taken by the recurrent laryngeal nerve on each side. In the early fetus, the vagus nerve lies lateral to the primitive pharynx, separated from it by the aortic arches. What are to become the recurrent laryngeal nerves pass medially, caudal to the aortic arches, to supply the developing larynx. With elongation of the neck and caudal migration of the heart, the recurrent nerves are caught up and dragged down by the descending aortic arches. On the right side, the 5th arch and distal part of the 6th arch are resorbed, leaving the nerve to hook round the 4th arch, i.e. the right subclavian artery. On the left side, the nerve remains looped around the persisting distal part of the 6th arch (the ligamentum arteriosum), which is overlapped and dwarfed by the arch of the aorta (see Fig. 82a).

The fetal circulation (Fig. 80)

The circulation of the blood in the embryo is a remarkable example of economy in nature that results in the shunting of well-oxygenated blood from the placenta to the brain and heart, leaving relatively desaturated blood for less essential structures.

Oxygenated blood is returned from the placenta by the umbilical vein to the inferior vena cava and thence to the right atrium, most of it bypassing the liver in the ductus venosus. Relatively little mixing of oxygenated and deoxygenated blood occurs in the right atrium, since the valve overlying the orifice of the inferior vena cava serves to direct the flow of oxygenated blood from that vessel through the foramen ovale into the left atrium, while the deoxygenated stream from the superior vena cava is directed through the tricuspid valve into the right ventricle. From the left atrium, the oxygenated blood (together with a small amount of

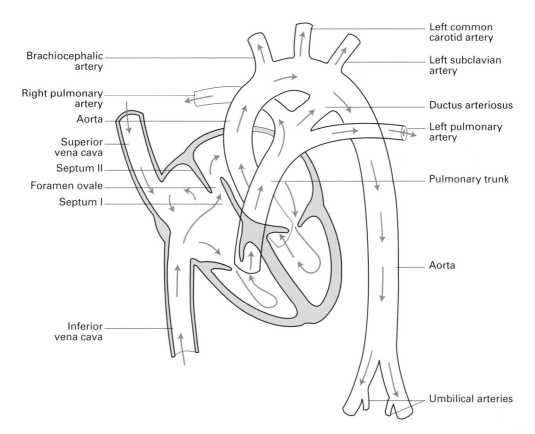

Fig. 80 The fetal circulation.

deoxygenated blood from the lungs) passes into the left ventricle and hence into the ascending aorta for the supply of the brain and heart via the vertebral, carotid and coronary arteries.

As the lungs of the fetus are inactive, most of the deoxygenated blood from the right ventricle is short-circuited by way of the ductus arteriosus from the pulmonary trunk into the descending aorta. This blood supplies the abdominal viscera and the lower limbs and is shunted to the placenta, for oxygenation, along the umbilical arteries, which arise from the internal iliac arteries.

At birth, expansion of the lungs leads to an increased blood flow in the pulmonary arteries. The uncoiling of the fetal pulmonary blood vessels results in a sudden, considerable reduction of pulmonary vascular resistance. The resistance to flow through the pulmonary artery decreases, whereas that of the systemic circulation increases. This results in a decrease in pressure in the right atrium, with an increase in the pressure within the left atrium. The resulting pressure changes in the two atria bring the septum primum and septum secundum into apposition and effectively close off the foramen ovale. At the same time, active contraction of the muscular wall of the ductus arteriosus as a result of the increased oxygen tension of the blood results in a functional closure of this arterial shunt and, in the course of the next 2–3 months, its complete obliteration. Similarly, division of the umbilical cord is followed by thrombosis and obliteration of the umbilical vessels after 3–4 weeks, although immediately after birth these vessels are readily available for transfusion.

Congenital abnormalities of the heart and great vessels

The complex development of the heart and major arteries accounts for the multitude of congenital abnormalities that may affect these structures, either alone or in combination.

Dextroposition of the heart means that this organ and its emerging vessels lie as a mirror image to the normal anatomy; it may be associated with reversal of all the intra-abdominal organs (situs inversus).

Septal defects include a *persistent patent foramen ovale* (which occurs in some 10% of subjects) and *atrial* or *ventricular septal defects*. An ostium secundum defect lies high up in the atrial wall and is relatively easy to close surgically. An ostium primum defect lies immediately above the atrioventricular boundary and may be associated with a defect of the pars membranacea septi of the ventricular septum; it thus presents a more serious surgical problem.

Occasionally, the ventricular septal defect is so huge that the ventricles form a single cavity, giving a trilocular heart.

Congenital pulmonary stenosis may affect the trunk of the pulmonary artery, its valve or the infundibulum of the right ventricle. If stenosis occurs in conjunction with a septal defect, the compensatory hypertrophy of the right ventricle (developed to force blood through the pulmonary obstruction) produces a sufficiently high pressure to shunt blood through the

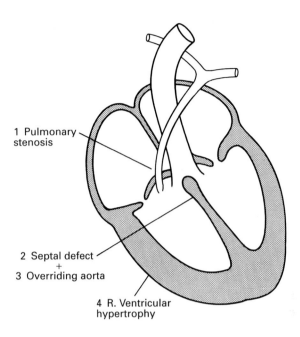

1 Pulmonary
stenosis

2 Septal defect
+
3 Overriding aorta

4 R. Ventricular
hypertrophy

Fig. 81 The tetralogy
of Fallot.

defect into the left heart; this mixing of the deoxygenated right heart blood
with the oxygenated left-sided blood results in the child being cyanosed
at birth.

The most common combination of congenital abnormalities causing
cyanosis is *Fallot's tetralogy* (Fig. 81). This results from unequal division
of the truncus arteriosus by the spiral septum, resulting in a stenosed
pulmonary trunk and a wide aorta that overrides the orifices of both the
ventricles. The displaced septum is unable to close the interventricular septum, which results in a ventricular septal defect. Right ventricular hypertrophy develops as a consequence of the pulmonary stenosis. Cyanosis
results from the shunting of large amounts of desaturated blood from the
right ventricle through the ventricular septal defect into the left ventricle
and also directly into the aorta.

A *patent ductus arteriosus* (Fig. 82a) is a relatively common congenital
defect. If left uncorrected, it causes progressive hypertrophy of the left
heart and pulmonary hypertension.

Aortic coarctation (Fig. 82b) is thought to be due to an abnormality of the
obliterative process that normally occludes the ductus arteriosus. There
may be an extensive obstruction of the aorta from the left subclavian artery
to the ductus, which is widely patent and maintains the circulation to the
lower part of the body; often, there are multiple other defects and frequently infants so afflicted die at an early age. More commonly, there is
a short segment involved in the region of the ligamentum arteriosum or
still-patent ductus. In these cases, circulation to the lower limb is maintained via collateral arteries around the scapula anastomosing with the
intercostal arteries, and via the link-up between the internal thoracic and

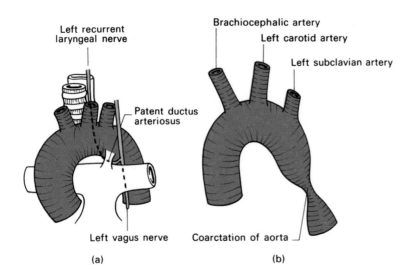

Fig. 82 (a) Patent ductus arteriosus showing its close relationship of the left recurrent laryngeal nerve; (b) coarctation of the aorta.

inferior epigastric arteries. Clinically, this circulation may be manifest by enlarged vessels that may be palpable around the scapular margins; radiologically, hypertrophy of the engorged intercostal arteries results in notching of the inferior borders of the ribs.

Abnormal development of the primitive aortic arches may result in the aortic arch being on the right, or actually being double. An abnormal right subclavian artery may arise from the dorsal aorta and pass behind the oesophagus – a rare cause of difficulty in swallowing (dysphagia lusoria).

Rarely, the division of the truncus into the aorta and pulmonary artery is incomplete, leaving an *aortopulmonary window,* the most unusual congenital fistula between the two sides of the heart.

The great veins of the neck

CLINICAL NOTE

Central venous pressure measurement, total parenteral nutrition and the use of percutaneous interventional cardiology techniques such as pacemakers have all made catheterization of the great veins of the neck and superior mediastinum an everyday hospital procedure. A detailed knowledge of the internal jugular, external jugular, subclavian and brachiocephalic veins is therefore of considerable importance.

The *internal jugular vein* runs from its origin at the jugular foramen in the skull (where it continues the sigmoid sinus) to its termination behind the sternal extremity of the clavicle, where it joins the subclavian vein to form the brachiocephalic vein (Fig. 83).

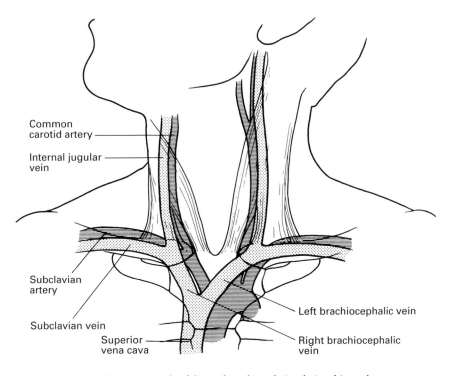

Fig. 83 The great vessels of the neck to show their relationship to the sternocleidomastoid and thoracic inlet.

It lies lateral first to the internal and then to the common carotid artery within the carotid sheath. In its upper part, the vein lies quite superficially in the anterior triangle of the neck, superficial to the external carotid artery, whose pulsations are usually visible as well as palpable. It then descends deep to the sternocleidomastoid muscle. The deep cervical chain of lymph nodes lies close against the vein and, if involved by malignant or inflammatory disease, may become densely adherent to it. The vagus lies between and rather behind the artery and vein. The cervical sympathetic chain ascends immediately posterior to the carotid sheath and these four structures, the two vessels and the two nerves, form a quartet that should all be considered in this inseparable manner; the relations of any one are those of the other three (Fig. 37, and see Fig. 211).

The following *tributaries* drain directly into the internal jugular vein:

1 the pharyngeal venous plexus;
2 the common facial vein;
3 the lingual vein;
4 the superior and middle thyroid veins.

The arrangement of the superficial veins of the head and neck is somewhat variable but the usual plan is as follows (Fig. 84): the *superficial temporal* and the *maxillary veins* join to form the retromandibular vein, which branches while traversing the parotid gland; the posterior division

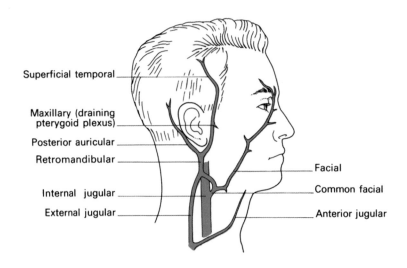

Superficial temporal

Maxillary (draining
pterygoid plexus)

Posterior auricular

Retromandibular

Internal jugular

External jugular

Facial

Common facial

Anterior jugular

Fig. 84 The usual
arrangement of the
veins of the neck.

continues to form the *external jugular vein*, whereas the anterior division joins the facial vein to form the *common facial vein*, which opens directly into the internal jugular.

The *external jugular vein* crosses the sternocleidomastoid in the superficial fascia, traverses the roof of the posterior triangle of the neck, then plunges through the deep fascia 2.5 cm above the clavicle to drain into the subclavian vein. As it pierces the deep fascia the vein tends to be splinted open. If lacerated at this site, air is likely to be sucked into its lumen and to lead to air embolism.

The *anterior jugular vein* runs down one on either side of the midline of the neck and crosses the isthmus of the thyroid. Just above the sternum it communicates with its fellow across the midline, then passes outwards, deep to the sternocleidomastoid, to enter the external jugular vein.

The *subclavian vein* is the continuation of the axillary vein and extends from its commencement at the outer border of the 1st rib to the medial border of scalenus anterior, where it joins the internal jugular vein to form the brachiocephalic vein (Fig. 83). During its short course, it crosses, and lightly grooves, the superior surface of the 1st rib (see Fig. 216). It arches upwards and then passes medially, downwards and slightly forwards to its termination behind the sternoclavicular joint. On the left side, it receives the termination of the thoracic duct. Anteriorly, the vein is related to the clavicle and subclavius muscle.

The *brachiocephalic (innominate) veins* (Fig. 85) are formed behind the sternoclavicular joints by the junction of the internal jugular and subclavian veins. Each lies lateral to the common carotid artery in front of scalenus anterior. On each side, it receives tributaries that correspond to the branches of the first part of the subclavian artery, i.e. the vertebral, inferior thyroid and internal thoracic veins. The vessels are asymmetrical.

The *right brachiocephalic vein* is approximately 3 cm in length and descends vertically behind the right border of the manubrium. The right phrenic nerve descends along its lateral surface, separating it from the pleura.

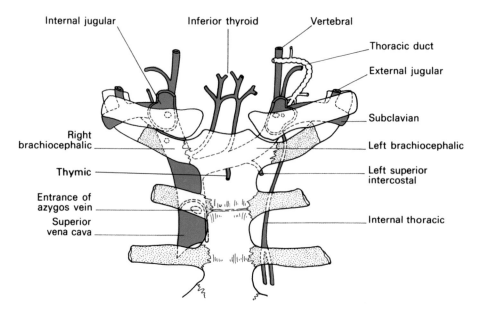

Internal jugular   Inferior thyroid   Vertebral

Thoracic duct

External jugular

Right brachiocephalic

Subclavian

Thymic

Left brachiocephalic

Left superior intercostal

Entrance of azygos vein

Superior vena cava

Internal thoracic

Fig. 85 The great veins of the neck and their tributaries.

The *left brachiocephalic vein* is approximately 6 cm in length and descends obliquely behind the manubrium to join the right vein at the lower border of the 1st costal cartilage to form the superior vena cava (Fig. 85). It crosses above the arch of the aorta in front of the left common carotid artery, the trachea and the brachiocephalic artery. As well as the tributaries common to both veins, the left brachiocephalic vein also receives the thymic vein and the left superior intercostal vein.

(Note that the dome of the pleura covered by the suprapleural membrane (Sibson's fascia) projects above the medial third of the clavicle for a distance of 2.5 cm. The convexity, which is well demonstrated on a chest radiograph, is due to the marked slope of the thoracic inlet. The internal jugular and brachiocephalic veins rest in close proximity to the pleura as their posterior relation (see Fig. 213). The risk of pleural puncture and therefore of a pneumothorax must be constantly borne in mind when needles are passed into this region.)

CLINICAL NOTE

Central venous catheterization
Central venous catheterization has become a routine procedure for high-risk surgery and intensive care and is commonly used on general wards and high-dependency units. This necessitates easy and safe access to the great veins. It facilitates central venous pressure monitoring, right heart catheterization, on occasion pulmonary artery catheterization (to measure cardiac output and mixed venous oxygen saturation; SVO_2), rapid blood transfusion and enables long-term intravenous feeding or drug therapy.

In the operating theatre and intensive care unit multilumen catheters are used. Central venous catheterization is most often achieved via the internal jugular and subclavian veins. It may be also achieved by means of a long cannula threaded from the antecubital fossa. However, it may be difficult to thread the catheter into the major veins as it may double-back on itself or become inadvertently threaded upwards into the internal jugular vein.

Subclavian vein and internal jugular vein catheterization carry complications and have a definite morbidity and even mortality, especially in inexperienced hands. Large haematomas may occur, especially after anticoagulation. Pneumothorax (3% when using a blind technique), haemothorax, air embolism, catheter embolism, thrombophlebitis, inadvertent cannulation of the right ventricle, puncture of a tracheal tube cuff, carotid arterial puncture (9.4% in blind techniques) and carotid vasospasm or embolus, arteriovenous fistula, cardiac tamponade from haemopericardium and torticollis, as a late result in infants, have all been reported. Partial dislodgement of the cannula may allow fluids and drugs to pass into the tissues of the neck or chest, and result in chemical damage to surrounding structures.

A major and most common complication is of catheter-related sepsis, which occurs in up to 10% of patients with mortality rates of around 20%. Because of this, full aseptic technique is mandatory for insertion (except in extreme circumstances).

For all these reasons, it is essential to be aware of the numerous possible complications of these procedures and to use the technique only when it is positively indicated. All catheters used are radio-opaque and, if the cannula is to be left *in situ*, its position should be checked radiographically. Many are antibiotic coated.

Imaging of central venous catheter insertion

It is recommended that two-dimensional ultrasound imaging is used for all central venous catheter insertions. Figure 86 shows a representation of

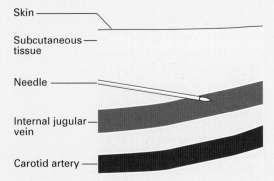

Fig. 86 Schematic representation of an ultrasonic view of internal jugular vein cannulation.

the image generated that allows for real-time localization of the needle tip and identification of surrounding structures. This device can be used for both subclavian vein and internal jugular vein cannulation. When using the right subclavian vein, a shorter catheter (15 cm) is often used to reduce the likelihood of right atrial cannulation. Venous blood is identified by its dark, non-pulsatile nature. If either route is used, the tip should be placed in the superior vena cava just at the point where it enters the right atrium.

Subclavian vein catheterization

The subclavian route for central venous cannulation is the preferred route for long-term tunnelled lines such as Hickman lines. There is a lower risk of infection than alternative routes such as the internal jugular vein and especially the femoral vein; the technique does not require the patient's head to be turned to the side; the site is more comfortable for the patient and easier to maintain in place if the line is to be left in for long periods. Evidence suggests that using the subclavian vein reduces other major complications. By far the most common technique is the infraclavicular approach. The needle entry point is at the junction of the middle and medial thirds of the clavicle, just inferior to it. The needle is aimed towards the suprasternal notch and is advanced under ultrasound imaging until blood is aspirated into the syringe. A wire is passed down the needle using a Seldinger technique and the catheter is advanced over the wire following dilation (Fig. 87). The subclavian vein is often found within 1–2 cm under the clavicle. However, the subclavian vein approach may be difficult as ultrasound location is not easy, and there is a greater risk of pneumothorax in patients whose lungs are mechanically ventilated; if the subclavian artery is accidentally cannulated it is not possible to apply pressure. It is said that

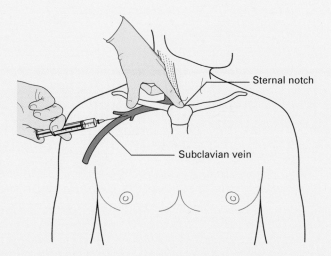

Fig. 87 Anatomical landmarks of subclavian puncture.

the left-sided subclavian approach can be associated with damage to the thoracic duct.

Internal jugular vein catheterization

The internal jugular route to central venous cannulation is used commonly for patients in intensive care units and patients who require short-term central line placement when the subclavian vein route is impractical. High and low approaches have been described, but the low approach is that most commonly used (Fig. 88). The patient is placed in the supine position, with the head to the contralateral side to the vein being catheterized. The patient should be tipped 10–20° head-down, if this is thought safe, in order to distend the veins of the neck. The mid-point of the sternocleidomastoid muscle, between its mastoid origin and its sternal insertion, is sought. The pulsation of the carotid artery can be felt medial to the medial border of the muscle. The vein lies lateral to the artery, often beneath the belly of the muscle itself, and its presence can often be appreciated by experienced fingers as a soft, 'fluid' feel. Lower in the neck, it passes deep to the groove between the sternal and clavicular heads of

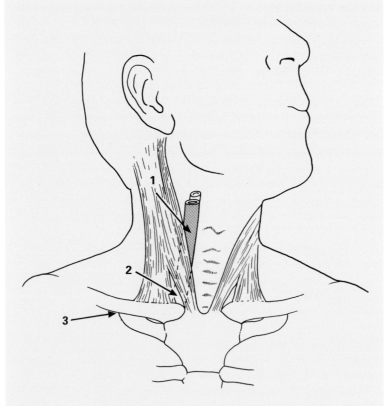

Fig. 88 Sites of cannulation of veins of the neck: 1, high internal jugular; 2, low internal jugular; 3, subclavian.

the muscle. Real-time ultrasound guidance is commonly used. A needle mounted on a syringe can be passed through the skin at an angle of approximately 30–40°, and classically aimed at the ipsilateral nipple, aspirating all the while to allow identification of correct placement in the vein. The vein is often found at a depth of 1–2 cm in adults. A wire is passed through the needle to facilitate catheterization in the same way as with subclavian vein cannulation.

Part 3
The Vertebral Canal and its Contents

The vertebrae and sacrum

The bones of the vertebral canal are landmarks, identified by both the palpating fingers and the exploring needle, by which the anaesthetist performs spinal and epidural blocks. As well as being able to recognize these landmarks, it is essential to be familiar with the feel of the intervertebral ligaments as they yield to the advancing needle, and to have an intimate knowledge of the relationship of nervous tissue and the dural sheath to the bony structures. Some anaesthetists have started using ultrasonic guidance and X-ray screening as an aid to spinal injections.

The vertebrae

There are seven cervical, 12 thoracic and five lumbar vertebrae. The sacrum comprises five, and the coccyx four, fused segments.

The adult spine presents four curvatures: those of the cervical and lumbar zones are convex forwards (lordosis); those of the thoracic and sacral regions are concave (kyphosis). The former are postural; the latter are produced by the actual configuration of the bones themselves. In the fetus, there is only a single concave-forward curvature; the cervical compensatory curve develops when the newborn infant holds up its head and the lumbar curve follows still later, when the child sits and then stands.

Although the individual vertebrae have their own features, they are constructed on a basic pattern as represented by the mid-thoracic vertebrae (Fig. 89): the *body*, through which the weight of the subject is transmitted, and the *vertebral* (or *neural*) arch, which surrounds and protects the spinal cord lying in the *vertebral foramen*. The arch comprises a *pedicle* and a *lamina* on each side, and a dorsal *spine*. Each lamina, in turn, carries a transverse process and superior and *inferior articular processes* that bear the *articular facets*. The pedicles are notched; the notches of each adjacent pair together form an intervertebral foramen through which emerges a spinal nerve.

The separate components of the spinal column will now be considered in more detail.

The cervical vertebrae

The typical cervical vertebrae are C3–6 (Fig. 90). Each of these has a small flattened body and a triangular, relatively large vertebral foramen. The pedicles project laterally as well as backwards, and their superior and inferior notches are about equal. The superior and inferior articular facets are on an articular pillar between the pedicle and the lamina, the superior facing upwards and backwards, and the inferior downwards and forwards. The transverse process is short (but readily palpable in a thin neck from the lateral aspect) and is pierced by the *foramen transversarium*, which transmits the vertebral vessels; it consists of an anterior and posterior root, each ending laterally in a tubercle, and connected to each other lateral to the foramen transversarium by the costotransverse bar. The anterior root and

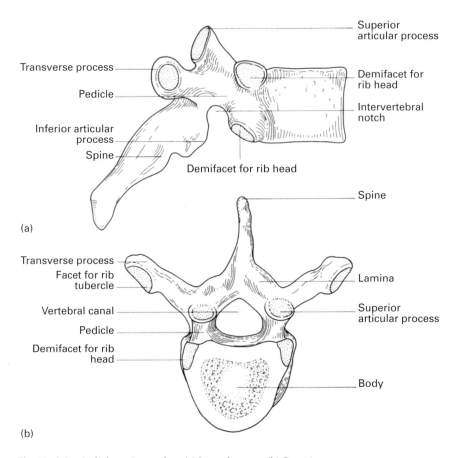

Fig. 89 A 'typical' thoracic vertebra. (a) Lateral aspect. (b) Superior aspect.

costotransverse bar are homologous with a rib and, like a rib, are attached to the side of the vertebral body. The anterior tubercle of C6 is large (Chassaignac's tubercle), it is palpable medial to the sternocleidomastoid, and against it can be compressed its anterior relation, the common carotid artery. The typical cervical spines are short and bifid.

The atlas, C1 (Fig. 91), has no body; instead, it consists of an anterior and a posterior arch joined by a thick lateral mass, which bears the superior and inferior articular facets and the transverse process (which is long and has no tubercles). The superior articular facets are strongly concave for articulation with the occipital condyles. The anterior arch bears a tubercle on its anterior aspect and a facet on its posterior surface, against which rests the dens (odontoid process) of C2. The upper surface of the posterior arch bears a deep broad groove just behind the superior articular facet. This groove is for the vertebral artery as this arches medially and upwards to enter the foramen magnum, and is also for the posterior primary ramus of the suboccipital nerve (C1), which emerges below the vertebral artery to enter the suboccipital triangle, where it supplies the adjacent muscles (Fig. 92). There is no spine; this is represented by a small posterior tubercle.

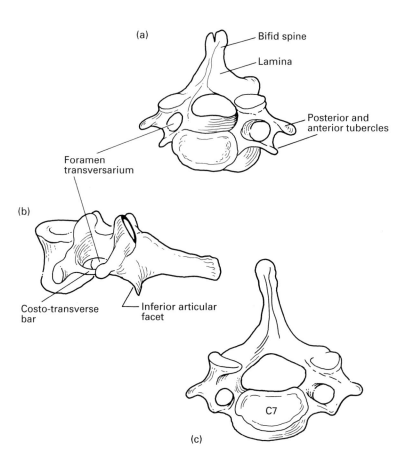

Fig. 90 Cervical vertebra in (a) superior and (b) lateral views. (c) The 7th cervical vertebra.

The axis, C2 (Fig. 93), bears its unmistakable *dens* or *odontoid process* on the superior aspect of its body. This process lies against the articular facet of the anterior arch of C1 separated from it by a small bursa. The superior articular facets are large, oval and face upwards and outwards; they are borne on the sides of the vertebral body. However, the inferior facets resemble those of the typical cervical vertebrae; they are carried on the laminae and face downwards and forwards. The transverse process is small and there is no differentiation into tubercles. The laminae are thick and the spine is large, strong and bifid.

The vertebra prominens, C7 (Fig. 90c), is so called because of its strong, non-bifid spine. On running the fingers down the nuchal furrow this spine is the first to be clearly palpable, although the spine of T1, immediately below it, is more prominent. The transverse process is large but its anterior tubercle is small and sometimes absent. The foramen transversarium is also small; this is because it transmits only accessory vertebral venules and not the vertebral artery, which enters at C6.

Note that the anterior primary rami of C3–7 issue anterior to the articular facets and thence pass behind the vertebral artery. However, C1 and C2 emerge behind their corresponding facets, and the anterior ramus of the suboccipital nerve (C1) passes forwards medial to the artery (Fig. 92).

(a)

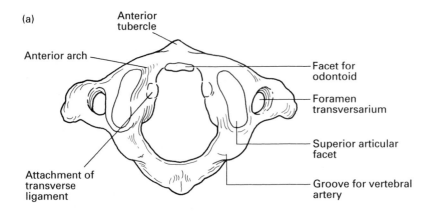

Anterior
tubercle

Anterior arch

Facet for
odontoid

Foramen
transversarium

Superior articular
facet

Attachment of
transverse
ligament

Groove for vertebral
artery

(b)

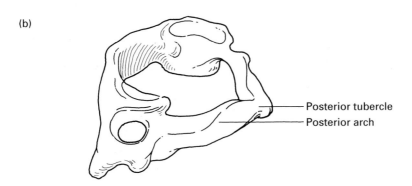

Posterior tubercle

Posterior arch

Fig. 91 The atlas in (a) superior and (b) oblique views.

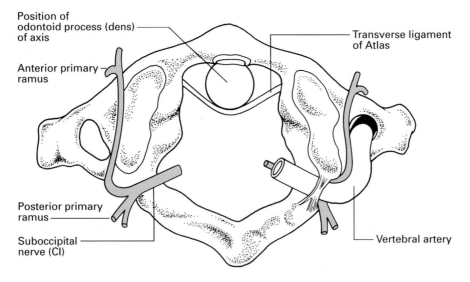

Position of
odontoid process (dens)
of axis

Transverse ligament
of Atlas

Anterior primary
ramus

Posterior primary
ramus

Suboccipital
nerve (CI)

Vertebral artery

Fig. 92 The relationships of the vertebral artery and the suboccipital nerve to the
atlas.

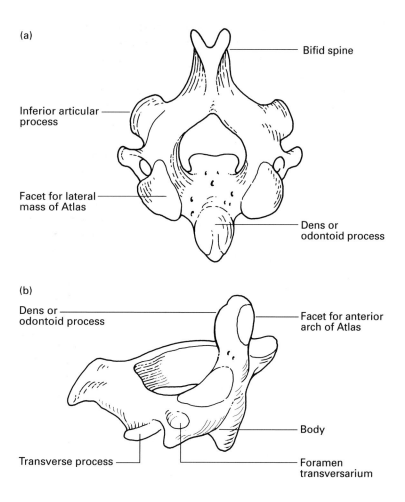

(a)

Bifid spine

Inferior articular process

Facet for lateral mass of Atlas

Dens or odontoid process

(b)

Dens or odontoid process

Facet for anterior arch of Atlas

Body

Transverse process

Foramen transversarium

Fig. 93 The axis in (a) superior and (b) oblique lateral views.

There are eight cervical nerves: C1–7 emerge above their corresponding vertebrae, but C8 lies above the first thoracic vertebra; the remaining spinal nerves emerge *below* their corresponding vertebra.

The thoracic vertebrae (Fig. 89)

The typical thoracic body is the conventional heart shape; the upper two bodies show a transition from the cervical type, whereas the lower vertebrae show some similarity to the lumbar bodies.

The bodies of T5–8 are flattened on their left side; this asymmetry is produced by the pressure of the descending aorta and it is these four vertebrae that become eroded by an aneurysm of this aortic segment. The typical vertebrae T2–8 bear upper and lower demifacets for the rib-heads: the superior demifacet is on the upper border of the side of the body at the base of the pedicle; the inferior is on the lower border, anterior to the inferior vertebral notch. The vertebral foramen is circular and its diameter is relatively small. The pedicles pass directly backwards, their superior vertebral notches, except for T1, are insignificant, but the inferior notches are deep. The laminae are broad and overlap each other from above down. The transverse processes are large, pass backwards and laterally, and typically bear a facet on the tip for articulation with the tubercle of the corresponding rib. The superior articular facets face backwards and outwards; the inferior facets correspondingly face forwards and inwards.

The spines are long. That of T1 projects almost horizontally backwards and is readily felt below the vertebra prominens; it is, in fact, the most readily palpable of the vertebral spines. The spines of the mid-thoracic vertebrae are angled caudally, and it is for this reason that, when performing a thoracic epidural with a midline approach, it is necessary to give a markedly cephalad angulation of the needle in order to pass between the spines. This is in contrast to the lumbar vertebral spines, whose upper borders are virtually at 90° both to the vertebral body and to the skin of the back. The spines of T11 and T12 are again nearly horizontal and are short, square and lumbar in type.

Features of the atypical thoracic vertebrae (Fig. 94)

The 1st thoracic vertebra has a cervical vertebral type of body, a marked upper notch, a complete upper facet for the 1st rib (since obviously there is no corresponding demifacet on C7) and a small demifacet below; its spine is horizontal.

The 9th thoracic vertebra is usually typical but often fails to articulate with the 10th rib-head; such specimens therefore will only possess a superior demifacet.

The 10th thoracic vertebra articulates only with the head of the 10th rib and therefore has only a superior demifacet (or a complete facet if there is no corresponding demifacet on T9). There may not be a facet on its transverse process.

The 11th thoracic vertebra articulates only with its own rib-head, for which there is a circular facet near the upper border of its body. The transverse process is small and facet-free.

The 12th thoracic vertebra has rather a lumbar-shaped body with a complete facet below its upper border. Its transverse process is small, without an articular facet but bearing superior, inferior and lateral tubercles like a lumbar vertebra. The inferior articular facet faces outwards; the spine is horizontal and resembles that of a lumbar vertebra.

The lumbar vertebrae (Fig. 95)

The bodies of the lumbar vertebrae are large and kidney-shaped; the vertebral foramen is roughly triangular, larger than in the thoracic but smaller than in the cervical region. The pedicles are thick, with shallow superior notches. The transverse processes are slender; they increase in length from L1 to L3, then become shorter again so that the third transverse process is longest; each bears an *accessory process* on the posteroinferior aspect of

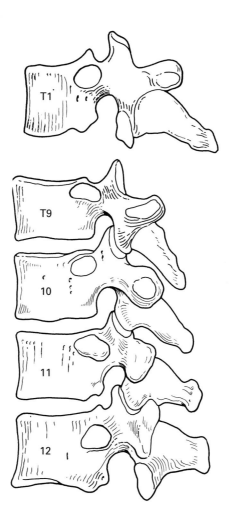

Fig. 94 The 'atypical' thoracic vertebrae in lateral view (T1, T9–12). The specimen of T9 shown has only a superior demifacet for the 9th rib-head.

its base, and a *mammillary process* adjacent to the superior articular process. The laminae are short, broad and strong, but they do not overlap each other as in the thoracic region. The superior articular facets face backwards and inwards; the inferior facets correspondingly face forwards and outwards. The lumbar spines are horizontal and oblong.

The 5th lumbar vertebra (Fig. 96), in producing the lumbosacral angle, is wedge-shaped, being considerably deeper in front than behind. Its transverse processes, although short, are thick and strong, and arise not only from the arch but also from the side of the vertebral body.

If the articulated vertebral column is inspected from behind, it will be noted that the laminae and spines so overlap and interdigitate with each other that the spinal canal is completely hidden, except in the lower lumbar region. This interlaminar gap is increased by forward flexion of the spine: a combination of circumstances that makes lumbar puncture possible (Fig. 97).

(a)

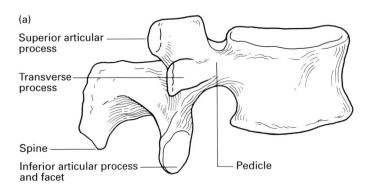

Superior articular process

Transverse process

Spine

Inferior articular process and facet

Pedicle

(b)

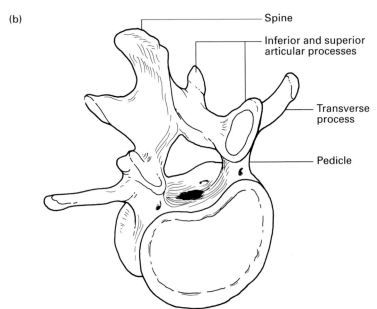

Spine

Inferior and superior articular processes

Transverse process

Pedicle

Fig. 95 Lumbar vertebra in (a) lateral and (b) anterosuperior views.

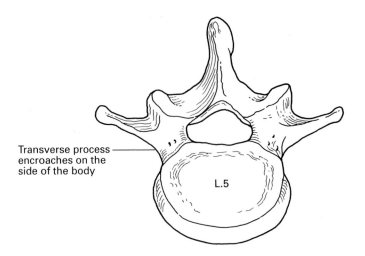

Transverse process encroaches on the side of the body

L.5

Fig. 96 The 5th lumbar vertebra, superior view.

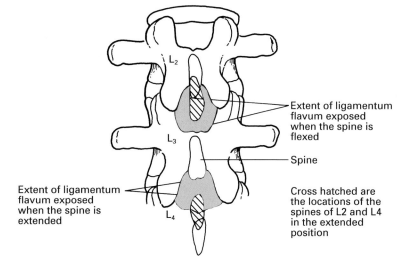

Fig. 97 The lumbar interlaminar gap; this anatomical fact makes lumbar puncture possible. Cross-hatched are the locations of the spines of L2 and L4 in the extended position.

CLINICAL NOTE

Spinal injections

A common technique for alleviation of chronic pain of spinal origin is the denervation of the facet joint by local anaesthetic medial branch block followed by radiofrequency denervation if satisfactory pain relief is produced (Fig. 98).

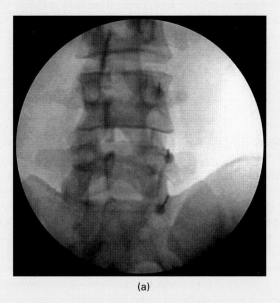

(a)

Fig. 98 (a) Anteroposterior screening image of needles for facet joint medial branch block. (b) Lateral screening image of facet joint medial branch blocks (see next page).

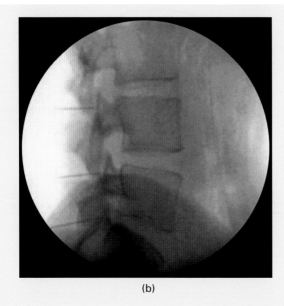

(b)

Fig. 98 (*Continued*)

CLINICAL NOTE

Lumbar puncture and spinal anaesthesia

Lumbar puncture (or spinal anaesthesia) is usually performed with the patient in the lateral or sitting position. Whichever position is chosen, the patient should be asked to flex his/her spine as much as possible, thereby widening the gaps between the lumbar spinous processes. The line that joins the top of the iliac crests (the intercristal line; Fig. 99) usually passes

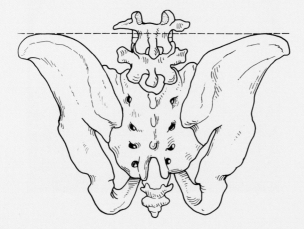

Fig. 99 A line joining the iliac crests (the intercristal line) is a good landmark for the identification of the 4th lumbar vertebra.

through the body of the 4th lumbar vertebra, and is therefore a useful land-mark. The space above this line is usually the L3/4 interspace; that below is usually the L4/5 interspace. The choice of interspace is important, as the spinal needle should not be introduced at a level that may cause it to enter the spinal cord. In the adult, the spinal cord usually ends at the level of the 1st lumbar vertebra. However, it may end as far distally as the 2nd or even 3rd lumbar vertebra (see Fig. 116). Spinal needles inserted for diagnostic or anaesthetic reasons should not therefore be introduced above the L3/4 interspace except in exceptional circumstances. Most anaesthesia for caesarean section in the UK is performed using a spinal technique (often in combination with intrathecal diamorphine).

Lumbar puncture is normally performed in the midline. After infiltration with local anaesthetic, the spinal needle is passed through the following structures (Fig. 100): skin; subcutaneous tissue; supraspinous and interspinous ligaments; ligamentum flavum; and dura mater. On puncturing the dura, a characteristic 'give' is often appreciated. On removal of the stylet from the needle, cerebrospinal fluid should appear at the hub of the needle.

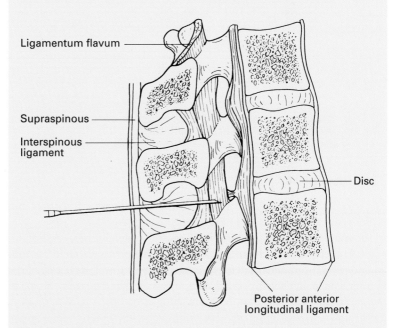

Ligamentum flavum

Supraspinous

Interspinous ligament

Disc

Posterior anterior longitudinal ligament

Fig. 100 The anatomy of lumbar puncture.

The lateral, or paramedian, approach to lumbar puncture, spinal anaesthesia or epidural anaesthesia is popular in some centres. It may be particularly useful in patients who have difficulty flexing their spines or in those whose supraspinous or interspinous ligaments are so calcified that passage of a needle through them proves difficult. It can also be used for the placement of high thoracic epidurals, as a steep cephalad angulation (as would be needed for a midline approach) is not necessary. Although different techniques exist, a common description is as follows. The needle entry point is some 1.5 cm lateral to the inferior border of a spinous process (Fig. 101). The needle is angled in a cephalad direction and slightly medially. If contact is made with a lamina, the needle is 'walked' in a cephalad direction until it passes through the ligamentum flavum. The epidural or subarachnoid space can then be accessed.

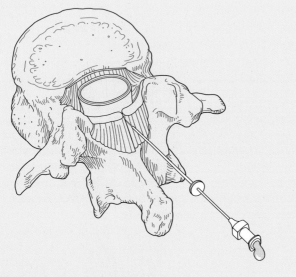

Fig. 101 The lateral approach to lumbar puncture.

The sacrum (Fig. 102)

The sacrum consists of five fused vertebrae. It is wedge-shaped and presents markedly concave anterior and convex posterior surfaces.

The anterior surface bears four transverse lines (demarcating the boundaries between the fused bodies), which terminate on each side in the four anterior sacral foramina, lateral to which is the fused lateral mass. The foramina lie in an almost parallel vertical row so that the wedge shape of the sacrum is due to the rapidly diminishing size of the lateral mass from above down. The anterior primary rami of the upper four sacral spinal nerves, as they emerge from the anterior foramina, produce distinct neural grooves on the lateral mass.

The posterior surface of the sacrum is made up of the fused vertebral arches that form the roof of the sacral canal. It presents a median crest of fused spines, represented by small spinous tubercles. On either side of this crest are the fused laminae, which bear laterally an articular crest composed of fused articular facets, each represented by a small tubercle; each articular crest terminates below in the *sacral cornu*. The last laminar arch (or more) is missing, leaving the *sacral hiatus*. Lateral to the articular tubercles are the four posterior sacral foramina, which lie directly opposite their corresponding anterior foramina and which are closed laterally by the

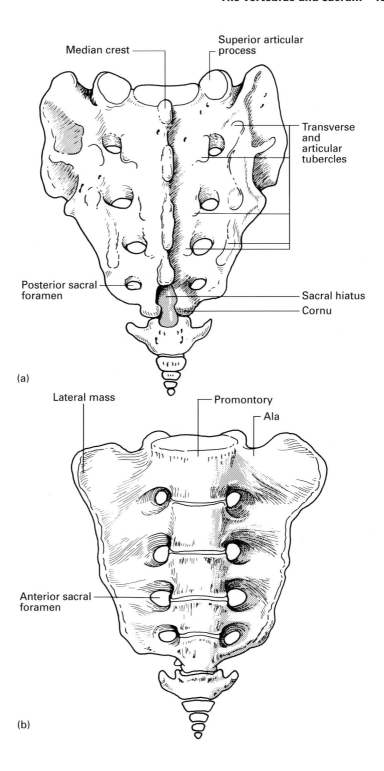

Fig. 102 The sacrum in (a) posterior and (b) anterior views.

posterior aspect of the lateral mass. The posterior rami of the sacral nerves emerge from the posterior sacral foramina. These nerves are small and relatively unimportant, and are concerned with the sensory innervation of a small area of skin over the sacrum and coccyx.

Lateral to these sacral foramina is a lateral sacral crest, formed by fused transverse processes, whose apices form the row of transverse tubercles. The posterior sacral foramina are continuous with the epidural space in the sacral canal.

CLINICAL NOTE

Injections of small volumes of local anaesthetic through these foramina will cause a segmental unilateral trans-sacral nerve block. Injection of local anaesthetic can produce temporary relief of pain resulting from irritation or compression of the sacral nerve roots. If the pain is caused by malignant disease, and temporary relief has been produced with local anaesthetic injections, a neurolytic block can be performed.

The posterior sacral foramina lie in a vertical line approximately 2 cm apart. The easiest foramen to detect is that of S2, which lies approximately 1 cm medial to and below the posterosuperior iliac spine, which itself lies deep to the sacral dimple and is thus easily identified. The foramen of S1 lies 1 cm above and medial to the dimple, and the foramina of S3 and S4 lie 2 and 4 cm vertically below the landmark of S2 (Fig. 103). Although the sacral foramina can be identified clinically, the use of an image intensifier is recommended for accurate location.

The lateral mass bears on its upper outer aspect a large *auricular surface* (which articulates with the corresponding auricular surface of the ilium),

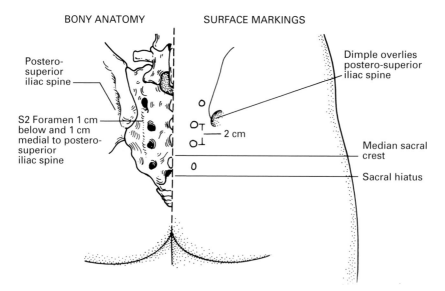

BONY ANATOMY   SURFACE MARKINGS

Postero-superior iliac spine

S2 Foramen 1 cm below and 1 cm medial to postero-superior iliac spine

Dimple overlies postero-superior iliac spine

2 cm

Median sacral crest

Sacral hiatus

Fig. 103 The surface markings of the posterior sacral foramina.

behind which is a large roughened area for the attachment of the strong posterior sacroiliac ligament.

The upper surface (or base) of the sacrum shows the features of a rather modified vertebra. The body is oval in section, its anterior edge forming the sacral promontory. The sacral canal is triangular in section, produced by very short pedicles and long laminae. The superior auricular facet faces backwards and inwards to receive the inferior facet of L5.

The upper surface of the lateral mass is termed the *ala* and is grooved by the lumbosacral cord (L4–5) of the sacral plexus.

The sacral hiatus, the triangular obliquely placed hiatus on the posterior aspect of the lower end of the sacrum, is of considerable practical importance; it is here that the epidural space terminates and the hiatus thus forms a convenient portal of entry into this compartment. The sacral hiatus results from failure of fusion of the laminae of the 5th sacral segment – or the defect may be more extensive than this and will be described under vertebral anomalies (see page 125). It is bounded above by the fused laminae of the 4th sacral segment (or of a still higher segment if the hiatus is more extensive), on which is situated the corresponding sacral spinous process. Laterally are placed the margins of the deficient laminae of S5, which below bear the sacral cornua, while inferiorly lies the posterior surface of the body of the 5th sacral segment.

The sacral hiatus usually lies 5 cm above the tip of the coccyx and directly beneath the uppermost limit of the natal cleft. In clinical practice, it is better to locate it by direct palpation of the depression which it forms between the sacral cornua. The hiatus is roofed over by the posterior sacrococcygeal ligament (approximately 1–3 mm thick), subcutaneous fat and skin; its ease of location varies inversely with the depth of the fat.

The coccyx consists of four fused rudimentary vertebrae, although the first often remains as a separate piece. This first segment carries poorly developed transverse processes and upper articular processes; the latter are termed the cornua of the coccyx.

CLINICAL NOTE

Caudal anaesthesia

The dural sac terminates at the level of the 2nd sacral vertebra (see Figs 105, 116). The epidural space continues below this point and can be accessed via the caudal (sacral) hiatus. Local anaesthetic injected at this level will primarily affect the sacral nerve roots, providing anaesthesia or analgesia for the perineum. Larger volumes of local anaesthetic will extend the effect to the lower lumbar roots. Paediatric anaesthetists occasionally use this route of access to the epidural space to pass an epidural catheter into the lumbar, or even thoracic, region. A technique which has also been described in adults.

The technique of caudal block depends upon identifying the caudal hiatus. The patient is placed in the lateral or prone position. The posterosuperior iliac spines and the caudal hiatus form an approximate equilateral triangle, allowing identification of the caudal hiatus. In slim

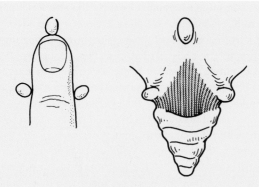

Fig. 104 The sacral cornua delimit the sacral hiatus.

patients, the sacral cornua can be appreciated as two adjacent knuckles approximately 5 cm above the tip of the coccyx at the upper end of the natal cleft (Fig. 104). A needle, often a 21G hypodermic needle, is introduced at an angle of approximately 45° to the skin, aiming to penetrate the posterior sacrococcygeal ligament and to enter the sacral canal. Once through the ligament, the needle hub is depressed so that the needle lies more parallel with the skin. The needle can then be advanced up the sacral canal before injection. As can be appreciated from Fig. 105, the needle can be placed incorrectly in a number of positions: subcutaneously; subperiosteally either within or outside the sacral canal; in an epidural vein; or within the dura if the needle is advanced too far. It is therefore important to aspirate before injection to exclude intravascular or subarachnoid placement and to confirm that injection is easily performed. Injection of radio-opaque dye is often used to confirm position when performed under image-intensifier guidance.

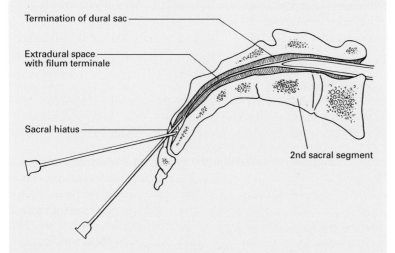

Termination of dural sac

Extradural space with filum terminale

Sacral hiatus

2nd sacral segment

Fig. 105 Longitudinal section through the sacrum to show the termination of the dura sac and the anatomy of caudal block.

Vertebral anomalies

Variations in the anatomy of the vertebral column are far from being only of academic importance to the anaesthetist, since they may render the performance of spinal or epidural blocks difficult or even impossible.

An appreciation of the developmental anomalies is simplified by a consideration of the embryology of this region. Mesodermal somites condense around the primitive notochord and neural tube. Each vertebral body originates from one half of each of two adjacent somites fusing together; the vertebrae are thus developmentally intersegmental. Primary centres of ossification develop one on either side of the vertebral arch and one within the body; occasionally, the latter comprises two centres side by side, which may fail to unite.

Defects in the vertebral bodies include the development of additional vertebrae or hemivertebrae – the latter producing a congenital scoliosis, anterior spina bifida, especially in the cervical and lumbar regions – owing to failure of the two centres in the body to unite, and absence either of vertebrae or of the lower sacrum and the coccyx (often associated with rectal and urinary incontinence). Fusion of two or more vertebrae may occur, and this may be complete or partial; this is particularly common in the sacral region, where the 5th lumbar vertebra may be wholly or partially fused to the sacrum (*sacralization of L5*). In contrast, the 1st sacral segment may be wholly or partially separated from the rest of the sacrum (*lumbarization of S1*).

Spina bifida

Neural arch defects result from the failure of fusion of the two arch centres, producing a spina bifida. Usually this is not associated with any neurological abnormality (spina bifida occulta), although in such cases there may be an overlying dimple, lipoma or tuft of hair to warn the observant of a bony anomaly beneath. More rarely, there is a gross defect of one or several arches with protrusion of the cord or its coverings.

The following range of anomalies may be found (Fig. 106):

1 *Spina bifida occulta* – failure of vertebral arch fusion only; the meninges and nervous tissue are normal.
2 *Meningocele* – protrusion of the meninges through a posterior vertebral defect without nervous tissue involvement.
3 *Myelomeningocele* – neural tissue (the spinal cord or roots) protrudes into, and may be adherent to, the meningeal sac.
4 *Myelocele* (rachischisis) – failure of fusion of the neural tube, producing an open spinal plate that occupies the defect as a red, granular area weeping cerebrospinal fluid (CSF) from its centre. This condition is incompatible with survival.

Spina bifida may occur anywhere along the vertebral column, but the great majority of defects involve L5 or the upper sacral region, with an incidence of 6% in L5 and 11% in the upper two sacral segments.

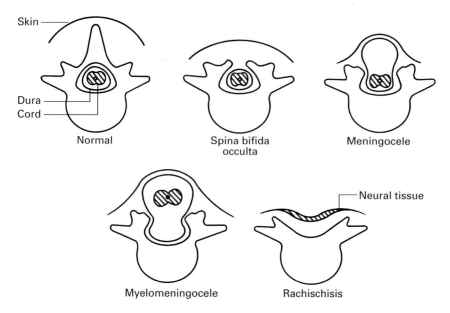

Fig. 106 The varieties of spina bifida.

Although the sacral hiatus classically reaches its apex at the level of the lower third of the body of the 4th sacral segment, this, in fact, is only true in 35% of humans. In 20% the apex is lower, even down to the lower border of S5, and in 45% it is higher, reaching the lower third of S2. Rarely, the sacral canal is open throughout its whole extent. Occasionally, there is a deficiency between the superior and inferior limits of the posterior wall of the sacral canal; there may then be a large defect either in the midline or to one side, or there may even be bilateral gaps on either side of the midline. Even with a normal hiatus, there may be deficiencies in the spines and laminae of the upper part of the sacrum.

The sacral canal may be wholly or partially obliterated by three anomalies. There may be a transverse fold in the posterior wall of the canal in conjunction with a forward projection of the corresponding segment of the sacral body, there may be a dorsal projection of a sacral vertebral body into the canal or there may be bony overgrowths that obliterate the hiatus.

Spondylolisthesis (Fig. 107)

A defect in the neural arch of L5 enables the whole spinal column, together with the body, pedicles and superior articular processes of L5, to slip forward on the sacrum, to which remain attached the laminae, spine and inferior articular processes of this vertebra. Occasionally, this defect affects L4.

The theory of the aetiology of the defect is that there may be two centres of ossification in the neural arch that fail to fuse; the two parts remain attached by fibrous tissue only and this may subsequently stretch, allowing the forward displacement to occur.

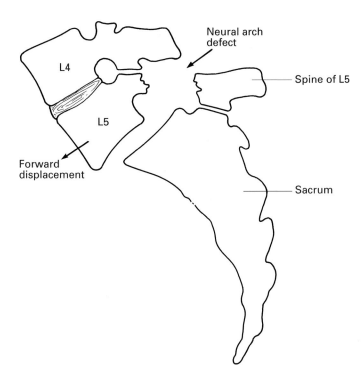

Fig. 107 The defect in spondylolisthesis.

The intervertebral ligaments (Fig. 108)

The individual vertebrae are linked to each other by a complicated system of intervertebral articular facets and ligaments. Only slight flexion, extension and rotation can take place between adjacent vertebrae, but these

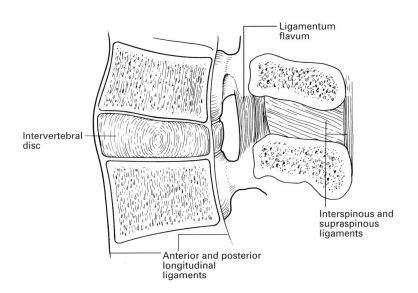

Fig. 108 Vertical section of the lumbar vertebrae to show the principal intervertebral ligaments.

small individual joint movements summate into the marked flexibility of the vertebral column as a whole.

Intervertebral cartilages (discs) are the principal connecting links between the vertebral bodies; they account for nearly 25% of the length of the spine. Each disc adheres above and below to the hyaline articular cartilage that covers the face of the adjacent vertebral body. In front and behind, each is attached to the anterior and posterior vertebral longitudinal ligaments. The disc is made up of peripheral fibrous tissue and fibrocartilage arranged in concentric rings, termed the *annulus fibrosus*, and a central core of soft, pulpy elastic tissue, the *nucleus pulposus*; this latter represents the remnant of the embryonic notochord. The discs are of uniform depth in the thoracic region, where the spinal curve is a function of the shapes of the vertebral bodies. In the cervical and lumbar regions, however, the cartilages are wedge-shaped (especially L5/S1), and contribute strongly to the curvatures in these zones. It is the atrophy of the intervertebral discs in the elderly, coupled with osteoporosis of the vertebrae, which accounts for both the shrinking height and kyphotic deformity of old age.

The *anterior longitudinal ligament* runs along the front of the vertebral bodies from C2 to the upper sacrum, becoming progressively wider from above downwards. It is adherent to the anterior aspect of each intervertebral disc and to the adjacent margins of the vertebral bodies.

The *posterior longitudinal ligament* extends along the posterior surfaces of the vertebral bodies. It corresponds in its attachments to those of the anterior ligament.

The *ligamenta flava* are ligaments of perpendicularly aligned elastic fibres that connect the adjacent laminae. They become progressively thicker from above downwards. For this reason, the elastic recoil that is transmitted to a syringe attached to an epidural needle inserted into the ligamentum flavum is more obvious in the lumbar than in the thoracic region. In elderly patients the elasticity of the ligamentum flavum tends to be lost and calcification of this structure may occur.

The *interspinous ligaments* connect the shafts of the adjacent spines; they are thin and tenuous, particularly in the cervical region.

The *supraspinous ligament* is a powerful column of fibrous tissue that connects the tips of the spines from C7 to the sacrum. This ligament may become ossified in elderly patients so that a midline spinal puncture becomes difficult.

The *ligamentum nuchae* consists of a superficial portion, representing the upward continuation of the supraspinous ligament, which stretches from C7 to the external occipital protuberance. From this superficial part, a fibrous sheet extends to become attached to the occipital bone and the spines of the cervical vertebrae.

The occipito-atlanto-axial ligaments (Fig. 109)

A complex series of joints link the skull and the upper two vertebrae; these are the articular facets between the occiput, atlas and axis, and the joint between the dens and the anterior arch of the atlas. We nod 'Yes' at the

(a)

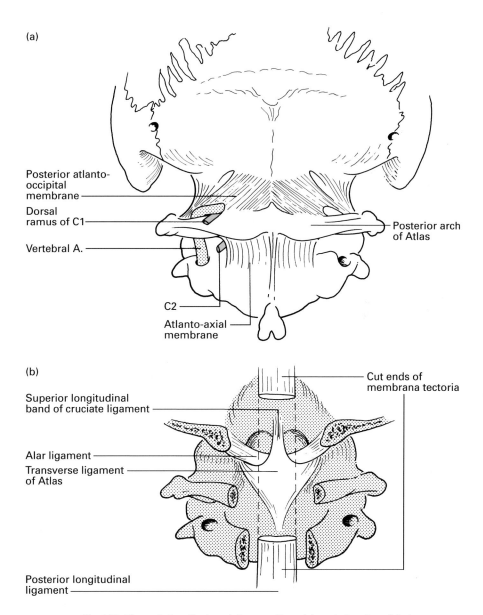

Posterior atlanto-occipital membrane

Dorsal ramus of C1

Vertebral A.

Posterior arch of Atlas

C2

Atlanto-axial membrane

(b)

Superior longitudinal band of cruciate ligament

Alar ligament

Transverse ligament of Atlas

Cut ends of membrana tectoria

Posterior longitudinal ligament

Fig. 109 The occipito-atlanto-axial connections: (a) posterior view; (b) view on removal of the laminae of the upper cervical vertebrae and the posterior part of the occipital bone.

atlanto-occipital, and shake 'No' at the atlanto-axial joint; the former also allows a degree of sideways tilting to take place.

A still more complex series of ligaments binds these joints together. Because of the wide range of movement allowed, the general rule for any highly mobile joint holds true – the ligaments close to the axis of rotation are short and strong, while those placed more peripherally, in this case linking the vertebral arches and the articular processes, are weak.

The posterior longitudinal spinal ligament is continued as the strong *membrana tectoria*, which stretches from the axis to the occiput, roofing over the atlas.

The anterior longitudinal ligament is continued upwards in the midline as a cord to the occiput. However, on either side it spreads out as the thin *anterior atlanto-occipital membrane*. The *transverse ligament of the atlas* stretches between the medial aspects of the lateral masses of this bone and holds the dens (odontoid peg) firmly against the articular facet on the posterior surface of the anterior arch (Fig. 92). Weak strands pass from this ligament upwards to the occiput and downwards to the body of the axis, so that the whole complex is termed the *cruciate ligament*.

The short and powerful *alar ligaments* pass from the sides of the apex of the odontoid peg to the medial sides of the occipital condyles; they serve to check excessive rotation of the skull. *Accessory alar ligaments* stretch on either side from the body of the axis to the lateral masses of the atlas. The *apical ligament* is a thin strand that links the tip of the odontoid with the anterior margin of the foramen magnum. It represents the remains of the upward prolongation of the notochord.

The *posterior atlanto-occipital membrane*, equivalent to the ligamentum flavum, extends from the posterior arch of the atlas to the occiput. It is pierced by the vertebral artery as this runs forwards and upwards to the foramen magnum.

The nuchal ligament has already been described (see above).

The sacrococcygeal connections

A thin fibrocartilaginous disc lies between the adjacent aspects of the sacrum and the coccyx. Linking their cornua is the *posterior sacrococcygeal ligament* that spreads out as a membrane covering the sacral hiatus and closing in the sacral canal.

The *anterior sacrococcygeal ligament* comprises a few weak longitudinal fibres that pass between the anterior aspects of the sacrum and coccyx.

The *lateral sacrococcygeal ligament*, on either side, connects the inferior lateral angle of the sacrum to the transverse process of the coccyx. It roofs over the 5th sacral nerve as this emerges between the cornua of the sacrum and coccyx.

The spinal meninges

The spinal cord has three covering membranes or meninges – the *dura* mater, the *arachnoid* mater and the *pia* mater (Fig. 110).

The dura mater

The dural covering of the brain is a double membrane, between the walls of which lie the cerebral venous sinuses. The dura mater that encloses the cord consists of a continuation of the inner (meningeal) layer of the cerebral

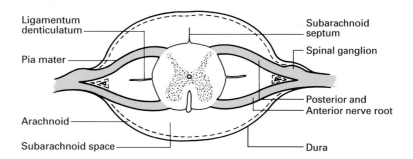

Ligamentum
denticulatum

Pia mater

Arachnoid

Subarachnoid space

Subarachnoid
septum

Spinal ganglion

Posterior and
Anterior nerve root

Dura

Fig. 110 The spinal
cord and meninges in
transverse section.

dura, which is made up of dense fibrous tissue; the outer (endosteal) layer
of the cerebral dura terminates at the foramen magnum, where it merges
with the periosteum enclosing the skull, and is thereafter represented by
the periosteal lining of the vertebral canal.

The dural sac usually extends to the level of the 2nd segment of the
sacrum; occasionally, it ends as high as L5; at other times it extends to S3.
As a result of this, it is occasionally possible to perform an inadvertent
spinal tap during the course of a caudal injection. The dural sheath then
continues as the covering of the filum terminale to end by adhering to
the periosteum on the back of the coccyx. The sac widens out in both
the cervical and lumbar regions, corresponding to the cervical and lumbar enlargements of the spinal cord. It lies rather loosely within the spinal
canal, buffered in the epidural fat, but it is attached at the following points
to its bony surroundings:

1 above, to the edges of the foramen magnum and to the posterior aspects
   of the bodies of the 2nd and 3rd cervical vertebrae;
2 anteriorly, by slender filaments of fibrous tissue to the posterior longitudinal ligament;
3 laterally, by prolongations along the dorsal and ventral nerve roots,
   which fuse into a common sheath and which then blend with the
   epineurium of the resultant spinal nerves;
4 inferiorly, by the filum terminale to the coccyx.
   However, the dural sac is completely free posteriorly.

The arachnoid mater

This is a delicate membrane that lines the dural sheath and that sends prolongations along each nerve root. Above, it is continuous with the cerebral
arachnoid, which loosely invests the brain and which dips into the longitudinal fissure between the cerebral hemispheres.

The pia mater

This, the innermost of the three membranes, is a vascular connective tissue sheath that closely invests the brain and spinal cord, and projects into
their sulci and fissures. The spinal pia is thickened anteriorly into the *linea
splendens* along the length of the anterior median fissure. On either side,

it forms the *ligamentum denticulatum*, a series of triangular fibrous strands attached at their apices to the dural sheath; they are 21 in number, and lie between the spinal nerves down to the gap between the 12th thoracic and 1st lumbar root. The lowermost denticulation is bifid and is crossed by the 1st lumbar nerve root.

The *posterior subarachnoid septum* consists of an incomplete sheet of pia passing from the posterior median sulcus of the cord backwards to the dura in the midline.

Inferiorly, the pia is continued downwards as the *filum terminale*, which pierces the lower end of the dural sac and then continues to the coccyx with a covering sheath of dura.

The compartments related to the spinal meninges

These are the subarachnoid, subdural and epidural spaces.

The *subarachnoid space* contains the CSF. It is traversed by incomplete trabeculae – the posterior subarachnoid septum and the ligamentum denticulatum, which have already been described (Fig. 110).

This space communicates with the tissue spaces around the vessels in the pia mater that accompany them as they penetrate into the cord. These continuations of the subarachnoid space have been described as breaking up into fine ramifications that surround individual nerve cells (the Virchow–Robin spaces) and that have been considered as pathways by which a spinal anaesthetic permeates the cord. It is debatable whether such spaces actually exist; they are probably artefacts produced by shrinkage in the course of fixation of the histological material.

The *subdural space* is a potential one only; the arachnoid is in close contact with the dural sheath and is separated from it only by a thin film of serous fluid. The subdural space within the vertebral canal rarely enters the consciousness of the clinician, unless it is the accidental site of catheter placement during attempted epidural analgesia or anaesthesia. The subdural injection of local anaesthetic is thought to be associated with patchy anaesthesia, often unilateral and often extensive.

The *epidural (extradural or peridural) space* (Fig. 111) in the spinal canal is that part not occupied by the dura and its contents. It extends from the foramen magnum to end by the fusion of its lining membranes at the sacrococcygeal membrane. It contains fat, nerve roots, blood vessels and lymphatics. The posterior aspect of the space is limited by the laminae and overlying ligamentum flavum, and at the sides by the pedicles of the vertebral arches and the intervertebral spaces. The front of the space is formed by the bodies of the vertebrae, the intervertebral discs and the posterior longitudinal ligament. The ligamentum flavum is 2–5 mm thick in cadavers and is divided by the vertebral spines into two parts, one arising from each lamina.

The segmental structure of the epidural space has been demonstrated using contrast radiography and magnetic resonance imaging (MRI). The wedge-shaped nature of each segment is especially marked in the midline.

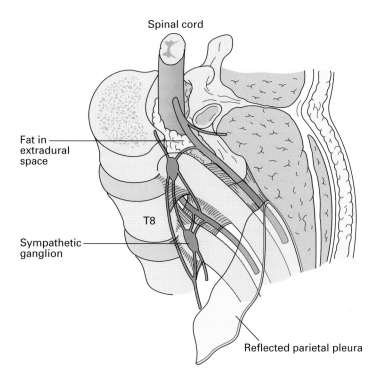

Fig. 111 The epidural (extradural) space.

Lateral radiographic studies following the injection of radiocontrast medium demonstrate a saw-tooth pattern of the epidural space. These studies suggest that the space is three to four times larger at the caudal than at the cephalic part of each segment. However, MRI studies have shown the lower end of the space in each segment to be approximately 4 mm wide compared with 6 mm at the upper end. This finding is in agreement with measurements made in cadavers. These differences illustrate the potential distortion due to the distensibility of this space after the injection of fluids. Some studies have suggested the presence of a posterior fold of dura termed the plica mediana dorsalis. It seems likely that this is an artefact caused by increased intrathecal pressure generated by the injection, as it is not seen on MRI. However, fibrous bands (sufficient to divert an epidural catheter) do stretch from the dura to the ligamentum flavum in a haphazard manner.

The capacity of the epidural space is far greater than that of the subarachnoid space at the same level. Thus, in the lumbar region, 1.5–2.0 ml of local anaesthetic is needed to block one spinal segment by the epidural route and only 0.3 ml to produce a similar extent of block by injection into the subarachnoid space. Each spinal nerve as it passes through its intervertebral foramen into the paravertebral space carries with it a collar of the fatty areolar tissue of the epidural space. Injection and dissection studies have shown that the paravertebral spaces, both serially and contralaterally, communicate with each other through the extradural space; there is, in fact, no direct communication between adjacent paravertebral spaces.

It is likely that there is a negative pressure in the epidural space. The communication between the epidural and paravertebral spaces explains this phenomenon. The paravertebral spaces in the thoracic region (Fig. 111) are only separated from the pleural cavities by the parietal pleura; pressure changes within the pleural cavity are thus transmitted to the paravertebral spaces in the thorax and thence to the epidural space. Deep breathing increases the negative epidural pressure but coughing will produce a positive pressure within this space. Pressure changes are most pronounced in the thoracic region but are elsewhere progressively dissipated by the buffering of the epidural fat, so that a negative pressure is no longer recorded within the cervical or sacral limits of the space. However, it has been argued that the negative pressure in the epidural space is produced, at least in part, by the tenting of the dura when a blunt epidural needle presses against it during insertion.

The epidural space can be entered by a needle passed either between the spinal laminae or via the sacral hiatus (see page 123). The spinal canal is roughly triangular in cross-section and, therefore, the space is deepest in the midline posteriorly. In the lumbar region, the distance between the laminae to the posterior aspect of the cord is approximately 5 mm. The distance from the skin to the lumbar epidural space varies between 2 cm and 7 cm, the range in the majority of patients being 3–5 cm.

The epidural space (Fig. 112) contains a network of veins. These run mainly in a vertical direction and form four main trunks: two lie on either side of the posterior longitudinal ligament and two posteriorly, in front of the vertebral arches. These trunks communicate freely by venous rings at each vertebral level. In addition, they receive the *basivertebral veins*, which

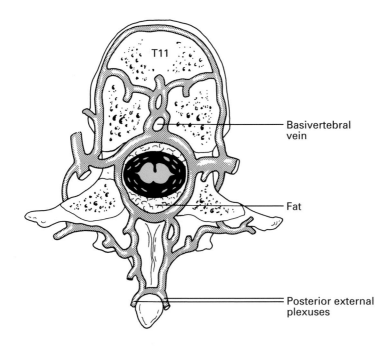

Fig. 112 Transverse section through T11 to show the epidural venous network. Note its immediate connections with the external vertebral venous plexuses.

emerge from each vertebral body on its posterior aspect, and communicating branches from the vertebral, ascending cervical, deep cervical, intercostal, lumbar, iliolumbar and lateral sacral veins that enter serially through the intervertebral and sacral foramina.

The epidural veins are valveless ('the valveless, vertebral, venous plexus of Bateson') and form a connecting link between the pelvic veins below and the cerebral veins above – a potential pathway for the spread of both bacteria and malignant cells.

The increase in CSF pressure that accompanies coughing and straining results in part from the shunt of blood from thoracic and abdominal veins into these thin-walled vertebral veins. The veins of the epidural space will therefore be distended if thoracic or abdominal pressure is increased, thus making a 'bloody tap' more likely.

The arteries of the epidural space are relatively insignificant and originate from the arteries corresponding to the named veins enumerated above. The arteries enter at each intervertebral foramen, lie chiefly in the lateral part of the epidural space and supply the adjacent vertebrae, ligaments and spinal cord.

CLINICAL NOTE

Epidural anaesthesia

It is common practice to use continuous epidural analgesia/anaesthesia in major abdominothoracic surgery, and in obstetric practice a combined spinal epidural technique is used when the procedure is expected to be prolonged. Using a Tuohy needle, the epidural space is accessed using a loss of resistance to saline technique. The needle is advanced through the ligamentum flavum with continuous pressure applied; as the aperture of the needle exits the ligamentum, the resistance to pressure is suddenly decreased. A fine-bore epidural catheter is then advanced into the epidural space. This technique can be used in any section of the spine (Fig. 113). A fine-bore spinal needle may be inserted through the Tuohy needle for combined techniques.

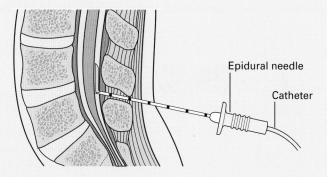

Epidural needle

Catheter

Fig. 113 Diagram of a lateral image of an epidural needle and catheter at L4/L5.

The cerebrospinal fluid

The CSF is the clear watery fluid contained within the cerebral ventricles and the subarachnoid space. The total volume of CSF in the adult is approximately 150 ml, of which some 25 ml is contained within the spinal theca.

CSF is produced by the choroid plexuses of the lateral, 3rd and 4th ventricles (Fig. 114); these plexuses are highly vascular invaginations of pia covered by ependymal epithelium that is but one cell thick. CSF is seen at operation oozing from the choroid plexus and unilateral hydrocephalus follows blockage of one foramen of Monro (leading from the lateral to the 3rd ventricle), but not if the choroid plexus is first removed; these three pieces of evidence are proof of the origin of the CSF. The CSF varies in composition from that of a true dialysate of plasma and, moreover, many soluble drugs fail to penetrate the 'blood–brain barrier' across the CSF; these points support the theory of active secretion and selective diffusion by the ependymal cells.

CSF escapes from the 4th ventricle through the median *foramen of Magendie* and the lateral *foramina of Luschka* into the cerebral subarachnoid space, at the cisterna cerebello-medullaris and cisterna pontis, respectively. About four-fifths of the fluid is reabsorbed via the *arachnoid villi*, minute projections of arachnoid that pierce the dural covering of the venous sinuses to lie immediately beneath their endothelium. Along the superior sagittal sinus, these villi clump together in adults to form the Pacchionian bodies, which produce the pitted erosions on either side of the median line of the inner aspect of the skull-cap.

The remaining one-fifth of the CSF is absorbed via similar spinal arachnoid villi or escapes along the nerve sheaths into the lymphatics. This absorption of CSF is passive, depending on its higher hydrostatic pressure than that of the venous blood.

There is a constant but slow circulation of CSF, which results in substances injected intrathecally being carried cephalad. Intrathecal injection

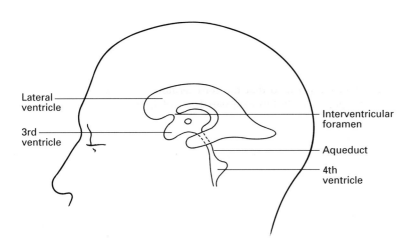

Lateral ventricle

3rd ventricle

Interventricular foramen

Aqueduct

4th ventricle

Fig. 114 The ventricular system.

of opioid drugs in the lumbar region may produce analgesia in the thoracic and cervical region and may cause late respiratory depression on reaching the medulla. This may occur up to 16 hours after the intrathecal injection.

Cerebrospinal fluid pressure

With the patient lying in the lateral position, the normal CSF pressure is approximately 70–180 mm of CSF. The dural theca acts as a simple valveless hydrostatic system so that, when the patient sits up, the pressure of the fluid in the lumbar part of the theca rises to between 350 and 550 mm of CSF, whereas the ventricular fluid pressure falls to below atmospheric.

Engorgement of the cerebral vertebral veins, as produced by coughing, straining and jugular venous compression (Queckenstedt's test), is transmitted to the buffering CSF and results in a brisk rise in CSF pressure. A part of the pressure increase in coughing and straining is due to a shunt of blood from the thoracic and abdominal veins into the spinal epidural venous sinuses (see page 135).

The spinal cord

The spinal cord is 45 cm long in the adult, a measurement it shares with the lengths of the femur and the vas deferens and with the distance from the lips to the oesophagogastric junction. It has an elongated cylindrical shape but is somewhat flattened anteroposteriorly, especially in the lumbar region. The cylinder is not uniform in diameter, but bears cervical and lumbar enlargements that correspond to the origins of the brachial and lumbosacral plexuses.

Below, the spinal cord tapers into the *conus medullaris*, from which a glistening thread, the *filum terminale*, continues down to become attached to the coccyx. The filum terminale is mainly pia mater invested in a sheath of dura, but it does contain a prolongation of the central canal of the cord in its upper part.

The relations of the cord to the vertebral column differ greatly in fetal, infant and adult life (Fig. 115). Up to the third fetal month, the cord extends the length of the vertebral canal. The vertebrae then grow considerably faster than the cord, so that the cord terminates in the newborn at the lower border of the 3rd lumbar vertebra and, in the adult, on average, at the disc between the 1st and 2nd lumbar vertebral bodies. However, there is considerable variation in this level (Fig. 116); frequently, the cord ends opposite the body of L1 or L2, or, rarely, T12 or even L3.

This differential growth results in the lumbar and sacral nerve roots becoming considerably elongated to reach their corresponding intervertebral foramina, thus forming the *cauda equina*. In contrast, the upper thoracic roots incline very little and the cervical roots pass almost laterally in their intraspinal course.

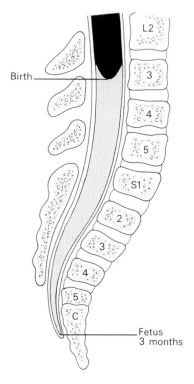

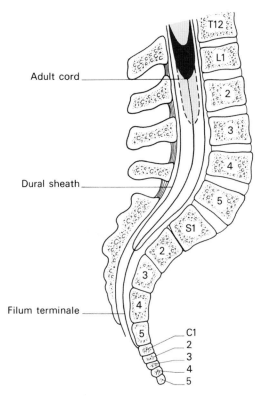

Fig. 115 The relationship between the spinal cord and the vertebrae in the 3 month fetus and in the newborn child.

Fig. 116 The range of variation in the termination of the spinal cord in the adult.

Table 1 Approximate levels of the cord segments

Cord segment	Vertebral body
C8	C7
T6	T4
T12	T9
L5	T12
Sacral	L1

Table 1 gives the approximate levels of the cord segments. As a rough guide, allow one segment difference in the cervical cord, two in the upper thoracic, three in the lower thoracic and four or five in the lumbar and sacral cord.

Following the injection of local anaesthetics or other fluids intrathecally, the spread will be influenced in part by the curvature of the spinal canal. When hyperbaric solutions are used, the local anaesthetic will tend to gravitate towards the lowest part of spine, i.e. to the bottom of the kyphoses. In the supine non-pregnant adult, these levels are L4 in the lumbar region and T8 in the thoracic region. In the pregnant patient, these levels are L4–5 and T6, the latter being, incidentally, an ideal dermatomal height for regional anaesthesia for caesarean section.

The structure of the cord (Fig. 117)

The spinal cord presents an *anterior median fissure* and a shallow *posterior median sulcus* from which a glial *posterior median septum* extends about

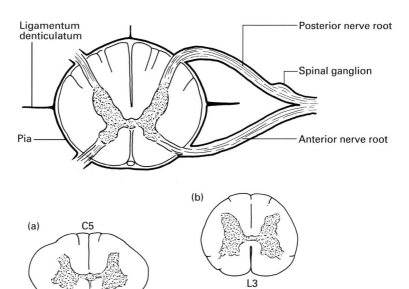

Fig. 117 The spinal cord in transverse section from a thoracic segment. Inset: (a) a section from the cervical cord; (b) a section from the lumbar cord.

halfway into the substance of the cord. On either side of the posterior sulcus lie the *posterolateral sulci*, along which can be seen the emerging line of posterior nerve roots. The anterior nerve roots, in contrast, emerge by a number of nerve tufts and their line of origin is not marked by a groove along the cord.

In transverse section the cord comprises a *central canal*, an 'H'-shaped zone of *grey matter* (nerve cells) and an outer zone of *white matter* (nerve fibres).

As the afferent nerve fibres are progressively added to the cord from below upwards and the efferent fibres are progressively given off from above downwards, the amount of white matter declines progressively from the cervical down to the lumbar region (Fig. 117). The grey matter is greatly increased in both the cervical and lumbar enlargements, which correspond to the zones of origin of the motor nerves to the upper and lower limbs.

The central canal continues downwards from the 4th ventricle as a narrow tube, lined with ciliated ependymal cells and containing CSF. It traverses the whole length of the cord, dilates somewhat within the conus medullaris, and continues for a short distance within the filum terminale.

The cross-limb of the 'H' of grey matter is termed the *transverse commissure*. Each lateral limb consists of a short, broad *anterior column*, containing large motor cells and a thinner, pointed *posterior column*, which is capped by the *substantia gelatinosa*. These columns are referred to, in descriptions of transverse sections of the cord, as the anterior and posterior horns, respectively. In the thoracic and uppermost lumbar segments lies the *lateral grey column*, which projects outwards from the grey matter at the junction of the anterior and posterior horns. It contains the spinal cells of origin of the sympathetic system.

The white matter consists, to a large extent, of longitudinally disposed medullated nerve fibres that can be divided, by their relationship to the central grey matter, into the posterior, lateral and anterior white columns. The two anterior columns are connected across the midline by the narrow white commissure immediately anterior to the grey commissure. The more important tracts in the white matter are as follows (Fig. 118).

Descending tracts

1 The *lateral cerebrospinal* or *pyramidal tract*, which is also termed the crossed motor tract, lies in the posterior part of the lateral white column and is the great motor pathway within the cord. This pathway commences in the pyramidal cells of the motor cortex of the cerebrum, decussates in the medulla (the pyramidal decussation) and then descends in the pyramidal tract on the contralateral side of the cord (Fig. 119). At each segment, fibres are given off that enter into synapse with the motor cells of the anterior horn, here forming the link-up between the upper and lower motor neurones. The pyramidal tract obviously becomes progressively smaller during its descent through the cord.

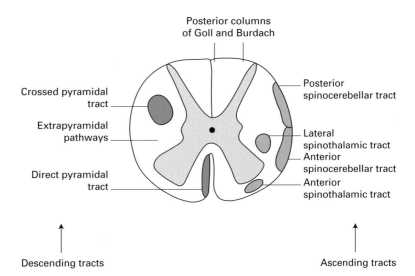

Fig. 118 The spinal tracts shown diagrammatically in a transverse section of the thoracic cord.

2 The *anterior cerebrospinal, direct pyramidal* or *uncrossed motor tract* is a small tract within the anterior white matter immediately adjacent to the anterior median fissure. The fibres in this tract descend from the motor cortex without medullary decussation. At each segment, however, fibres pass from it to the motor cells of the *opposite* anterior horn.

Ascending tracts

1 The *posterior column* of white matter is made up entirely of the medial *fasciculus gracilis* of Goll and lateral *fasciculus cuneatus* of Burdach; these convey sensory fibres subserving fine touch and proprioception (position sense), mostly uncrossed, to the gracile and cuneate nuclei in the medulla, respectively. Here, after synapse, the fibres cross in the medullary sensory decussation, pass to the thalamus in the medial lemniscus and are thence relayed to the sensory cortex (Fig. 120). Some fibres pass from the medulla to the cerebellum along the inferior cerebellar peduncle.

2 The *spinothalamic tracts*: pain and temperature fibres, together with some tactile afferent fibres, enter the posterior roots, ascend one or two segments, relay in the substantia gelatinosa, then cross the cord to ascend on the opposite side to the thalamus. Here they are relayed to the sensory cortex (Fig. 120). The pain and temperature fibres are contained in the lateral spinothalamic tract, within the lateral white substance anterior to the pyramidal tract. The tactile fibres lie within the anterior spinothalamic tract, immediately anterior to the tip of the anterior horn of the grey matter.

3 The *anterior* and *posterior spinocerebellar tracts* lie on the outer margin of the lateral white matter. Proprioceptive sensory fibres ascend through them, uncrossed, to the cerebellum, which they enter through the superior and inferior cerebellar peduncles, respectively.

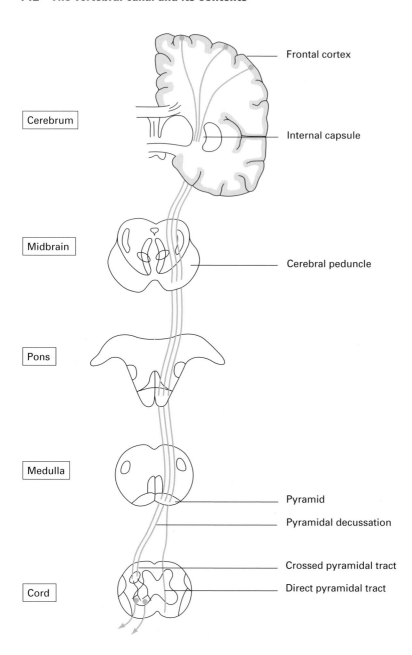

Fig. 119 The descending pathway of the pyramidal tract.

Blood supply

The arteries supplying the spinal cord are the anterior and posterior spinal arteries, which both descend from the level of the foramen magnum.

The *anterior spinal artery* is a midline vessel lying on the anterior median fissure and is formed at the foramen magnum by the union of a branch from each vertebral artery. It is the larger of the two vessels and supplies the whole of the cord in front of the posterior grey columns. However, its

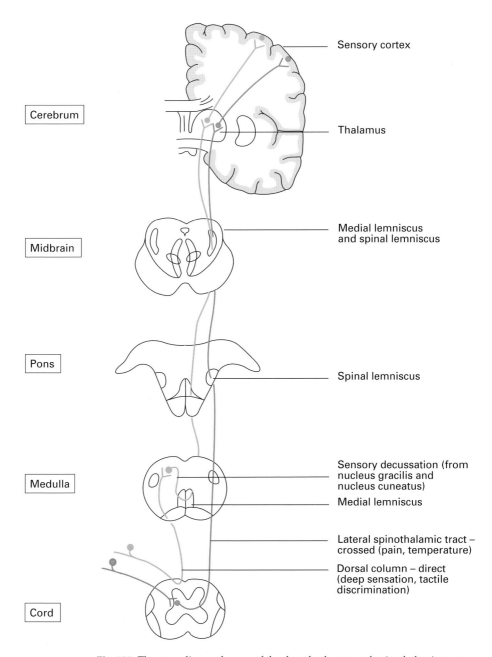

Fig. 120 The ascending pathways of the dorsal columns and spinothalamic tracts.

diameter may vary considerably as it passes down the cord and may, at times, disappear completely for short segments.

The *posterior spinal arteries* comprise one or two vessels on either side derived from the posterior inferior cerebellar arteries. They supply the posterior grey and white columns on either side. These arteries are reinforced serially by spinal branches of the vertebral, deep cervical, intercostal, lumbar, iliolumbar and lateral sacral arteries; the lower branches being responsible for the blood supply of the cauda equina. These vessels pass through the intervertebral foramina along the ventral and dorsal nerve roots, which they supply.

Most of the anterior radicular arteries are small, but a variable number (usually between four and nine), which are usually situated in the lower cervical, lower thoracic and upper lumbar regions, are larger and reinforce the anterior spinal arteries (Fig. 121a).

Often, one of the anterior radicular arteries is of considerable size and termed the *arteria radicularis magna*, or artery of Adamkiewicz. Its position is variable, but it usually arises in the lower thoracic or upper lumbar region (Fig. 121a). This is usually single and in the majority of cases arises on the left-hand side. On reaching the spinal cord, this artery sends a branch to join both the anterior and the posterior spinal artery and may be responsible for most of the blood supply of the lower two-thirds of the spinal cord.

The posterior radicular arteries are variable in number and size, and reinforce the posterior spinal artery together with the arteria radicularis magna. The anterior and posterior spinal arteries do not anastomose with each other within the spinal cord, and thrombosis of these vessels will result in cord infarction (Fig. 121b). The arterial supply to the spinal cord is very vulnerable and spontaneous occlusion of these vessels may occur from relatively trivial trauma. The cord blood supply may also be jeopardized in surgical cross-clamping of the thoracic aorta and dissection of the aorta. It may also result from the use of vasoconstrictors in epidural anaesthesia or may follow deliberate or accidental hypotension.

The venous drainage comprises a plexus of anterior and posterior spinal veins that drain along the nerve roots through the intervertebral foramina into the segmental veins: the vertebral veins in the neck, the azygos veins in the thorax, lumbar veins in the abdomen and lateral sacral veins in the pelvis. At the foramen magnum, they communicate with the medullary veins.

CLINICAL NOTE

Complete transection, hemisection and syringomelia
Complete transection of the cord is followed by total loss of sensation in the regions supplied by the cord segments below the level of injury together with initial flaccid muscle paralysis during the period of spinal shock. This is followed by spasticity of the paralysed lower limbs. Voluntary sphincter control is lost, but reflex emptying of rectum and bladder subsequently return, providing the cord centres in the sacral zone of the cord are not destroyed. Lesions high enough to involve the 4th cervical cord segment are usually fatal because of phrenic nerve paralysis.

Hemisection of the cord, from injury or disease, results in the Brown-Séquard syndrome. There is paralysis of the muscles innervated below the

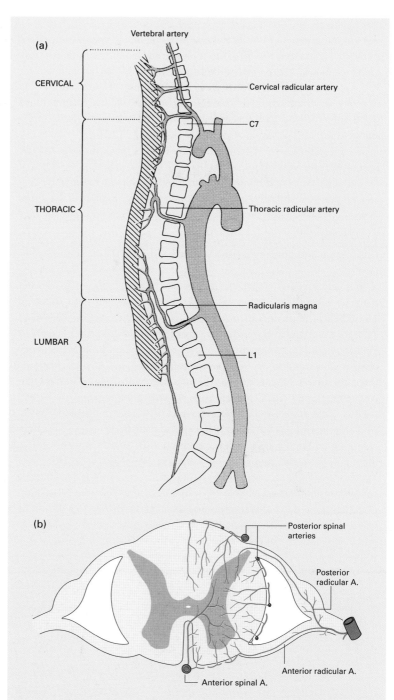

Fig. 121 The arterial supply of the spinal cord. (a) In schematic lateral view. (b) In transverse section.

level of the lesion and on the same side as the injury (transection of the pyramidal tract) and loss of position sense, movement sense and tactile discrimination in the paralysed limb (transection of the dorsal columns, which ascend uncrossed), but loss of pain and temperature sense in the opposite, unparalysed, limb because of division of the decussating fibres of the spinothalamic tract that then ascend in the contralateral side of the spinal cord.

Syringomyelia, a cystic degenerative condition of the centre of the upper part of the cord, results in the first instance in destruction of the fibres of the spinothalamic tracts on each side as these cross in the midline; the disease presents with bilateral loss of pain and temperature sense in the upper limbs, and consequently may result in severe scarring and tissue damage from unnoticed injuries.

Cordotomy and dorsal root entry zone lesion

Cordotomy, the surgical division of the lateral spinothalamic tract, may be used to relive severe, intractable and unilateral lower limb, pelvic or trunk pain. The most promising results seem to come in patients with unilateral cancer pain. This procedure obviously depends on the exact localization of the tract within the cord. Cordotomy is usually performed percutaneously, guided by computed tomography or image intensification. Radiofrequency electrodes are used to confirm the correct position and then to create the thermal lesion in the target area. The procedure aims to interrupt the lateral spinothalamic tract (carrying pain impulses from the opposite side of the body), but preserves the pyramidal tract, which lies immediately posterior to this line of section. Incomplete relief of pain, or its recurrence within weeks of cordotomy, may occur, as a proportion of afferent nociceptive input enters the cord via the ventral horn (Fig. 118).

Dorsal root entry zone (DREZ) lesion, in which, in a similar fashion to a cordotomy, a radiofrequency electrode may be inserted into the dorsal horn of the spinal cord to produce a thermal lesion of the DREZ. This technique has been tried for many chronic pain syndromes, but has been shown to be particularly effective in brachial plexus avulsion injuries.

Part 4
The Peripheral Nerves

Anatomy for Anaesthetists, Ninth Edition. Harold Ellis and Andrew Lawson.
© 2014 John Wiley & Sons, Ltd. Published 2014 by John Wiley & Sons, Ltd.

The spinal nerves

There are 31 pairs of spinal nerves – eight cervical, 12 thoracic, five lumbar, five sacral and one coccygeal. Each is formed by the fusion of an anterior and posterior spinal root.

The *anterior* (*ventral*) *roots* (Fig. 117) are motor and emerge in series from the anterior grey column of the spinal cord, each as a tuft of nerve rootlets.

The *posterior* (*dorsal*) *roots* (Fig. 117) are sensory and enter the cord in series along a posterolateral groove overlying the posterior grey column. Each posterior root carries a ganglion, immediately distal to which the anterior and posterior roots meet to form a spinal nerve.

In addition to the anterior and posterior nerve roots, the spinal cord bears a third and lateral set of nerve roots: the series of filaments from the upper four to six cervical segments that unite to form the spinal root of the accessory nerve (page 289). This root ascends alongside the cord through the foramen magnum.

The *spinal nerves* each give off a small meningeal branch, which re-enters the intervertebral canal and supplies the adjacent blood vessels and ligaments, then almost at once divide into the anterior and posterior primary rami.

The arrangement of a 'typical' spinal nerve (Fig. 122) is as follows:

1 The *posterior primary ramus* passes backwards between the transverse processes and then divides into a medial and lateral branch. These supply the adjacent vertebral muscles and send (from one or other branch) a cutaneous supply to the overlying skin.

2 The *anterior primary ramus* is linked to the sympathetic chain by a white and grey ramus communicans. It then runs in the body wall, and about half-way along its course gives off a lateral cutaneous branch that divides into anterior and posterior branches.

The nerve ends anteriorly by becoming the anterior cutaneous branch that supplies an area of skin adjacent to the midline. This 'typical' plan is seen only in the thoracic segments; elsewhere, it is modified, especially in relation to the formation of the great nerve plexuses from the anterior primary rami. These modifications are considered later in detail.

Meningeal relations

Both the anterior and posterior nerve roots bear prolongations of the pia and arachnoid mater that end where the roots separately pierce the dura. A cuff of dura then covers the roots as far as their junction to form the spinal nerve; here, the dura merges into the nerve sheath.

Vertebral relations

The posterior root ganglia lie in the intervertebral foramina. The only exceptions to this rule are those of C1 and C2, which lie on the posterior

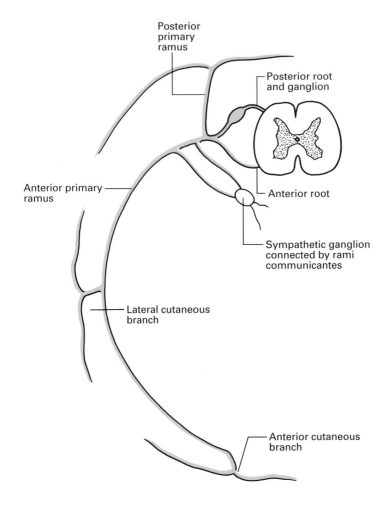

Posterior
primary
ramus

Posterior root
and ganglion

Anterior primary
ramus

Anterior root

Sympathetic ganglion
connected by rami
communicantes

Lateral cutaneous
branch

Anterior cutaneous
branch

Fig. 122 Plan of the
arrangement of a
typical spinal nerve.

arches of their respective vertebrae, and the sacral and coccygeal ganglia,
which remain within the sacral canal.

The 1st cervical nerve emerges between the occiput and the posterior
arch of the atlas; it is appropriately termed the suboccipital nerve. Nerves
C2–7 similarly emerge above their respective vertebrae, but C8 passes
through the intervertebral canal between the 7th cervical and 1st thoracic
vertebra. Below this level each spinal nerve emerges *below* its corresponding vertebra.

Since the adult spinal cord ends, on average, at the level of the disc space
between the 1st and 2nd lumbar bodies, the anterior and posterior nerve
roots become increasingly elongated from above downwards to reach their
intervertebral foramina (page 137).

The paravertebral space

A potential space exists along the outside of the vertebral canal, which
constitutes the paravertebral space. On the medial side lie the pedicles and
intervertebral foramina, and on the lateral side the pleura and lungs in the

thoracic region, and the sacrospinalis muscle supplemented by the origins of the psoas muscle in the lumbar region. The space is filled with loose fat and areolar tissue and contains the nerve roots issuing from the spinal cord together with blood vessels. This area is of interest to anaesthetists as it allows access for blocking the nerve roots without invading the epidural or subarachnoid space.

CLINICAL NOTE

By injecting fluid containing local anaesthetic into this space, unilateral nerve blocks extending over two to four segments can be obtained. This is of use in the cervical region to provide analgesia of the upper limb, in the thoracic region to produce pain relief for thoracic, breast, renal and biliary surgery, and in the lumbar region to produce analgesia for lower abdominal, hip and leg surgery. Identification of the paravertebral space can be achieved by loss of resistance, with a nerve stimulator, seeking muscle contractions within the distribution of the target nerve root and/or combined with ultrasound guidance. Paravertebral catheterization is increasing in popularity, allowing prolonged continuous paravertebral analgesia after surgery.

The posterior primary rami

The posterior primary rami of the spinal nerves are concerned with the innervation of the skin and muscles of the back. Those of C1 and C2 are quite exceptional, but certain generalizations can be made about the remainder as a preamble to their more detailed consideration.

1 The posterior primary rami supply motor and sensory fibres to serially segmental areas that slope downwards and outwards from the corresponding vertebral level. These segments overlap, so that surgical division or anaesthetic block of one single nerve does not reliably produce a corresponding band of cutaneous anaesthesia.

2 Unlike the anterior primary rami, the posterior rami do not extend into either the upper or the lower limb and do not form plexuses. With the exception of C1 and C2, each posterior primary ramus is smaller than its corresponding anterior ramus; a fact that can be correlated with the smaller demands made upon the former.

3 With the exception of C1, S4, S5 and Co. (coccygeal) 1, each posterior primary ramus divides into a medial and a lateral branch within the dorsal muscle mass. The cutaneous component is contained, in segments T6 and above, in the medial branch, but below T6 is transmitted in the lateral branch. No cutaneous fibres at all are conveyed by C1, C6–8 or L4–5 (Fig. 123).

The cervical posterior primary rami

The posterior primary ramus of C1 is larger than the corresponding anterior primary ramus, is entirely motor and does not divide into lateral

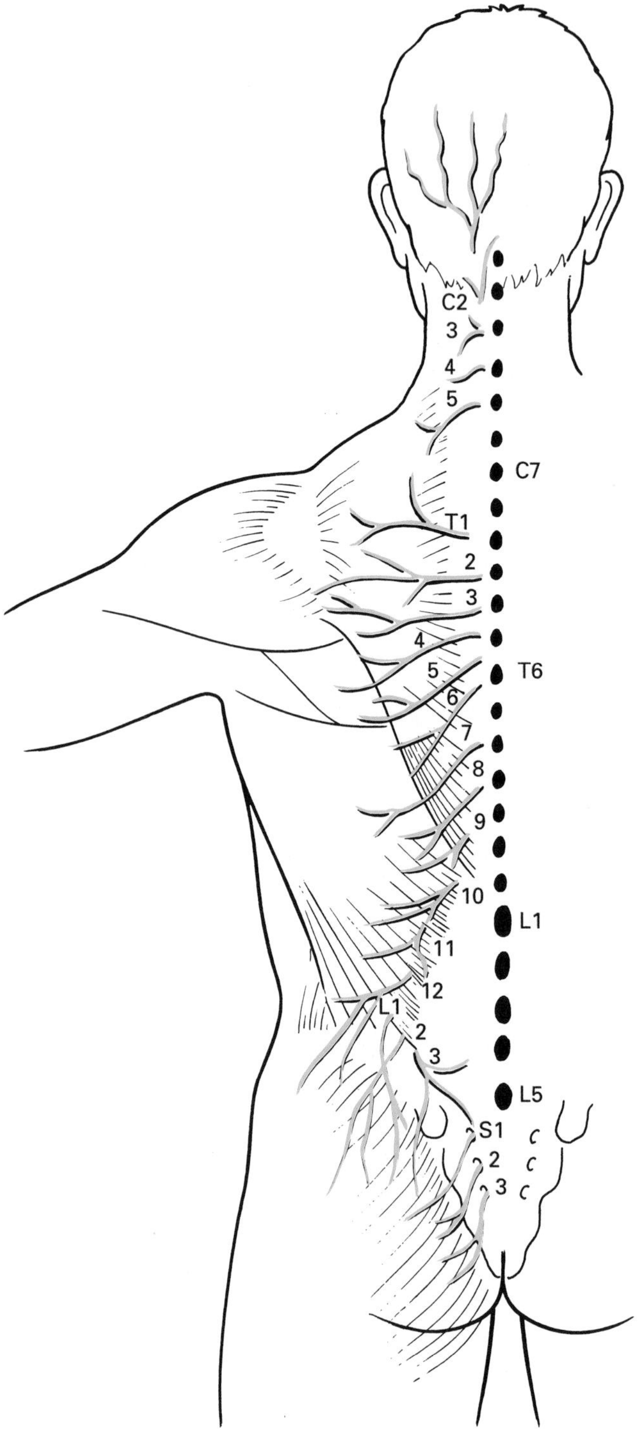

Fig. 123 The distribution of the cutaneous branches of the posterior primary rami.

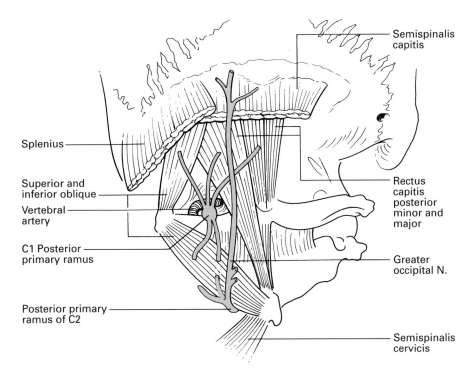

Fig. 124 The suboccipital triangle.

and medial branches. It emerges over the posterior arch of the atlas, sandwiched between bone and the vertebral artery (Figs 92, 124), to enter the suboccipital triangle, where it supplies the three surrounding muscles: the superior oblique, inferior oblique and rectus capitis posterior major. In addition, branches pass to rectus capitis posterior minor and to semispinalis capitis.

The posterior primary ramus of C2 is again larger than its anterior primary ramus and is, in fact, the largest of the cervical posterior primary rami. It emerges between the posterior arch of the atlas and the lamina of the axis, curves round the inferior border of the inferior oblique muscle, to which it sends a branch (Fig. 124), and then divides into a large medial and a small lateral branch.

The medial branch is the *greater occipital nerve*, which pierces semispinalis capitis and then trapezius, is joined by a filament from the medial branch of C3 and then ascends medial to the occipital artery to supply the skin of the occipital region as far as the vertex. Anteriorly, it overlaps with the lesser occipital nerve, derived from the anterior primary ramus of C2. It gives a branch to semispinalis (see Fig. 127).

This nerve is sometimes involved in cervicogenic headache. Infiltration of the greater occipital nerve with local anaesthetic may produce analgesia. However, it must be borne in mind that it is possible for C2 pain to be referred to areas supplied by the trigeminal nerve. The nerve is blocked by injecting local anaesthetic just medial to the occipital artery at a point one-third of the distance between the greater occipital prominence and the mastoid process.

The lateral branch is entirely motor; it partly supplies the posterior cervical muscles.

The medial branch of the cervical posterior primary ramus of C3 constitutes the *third occipital nerve*; it supplies the skin of the lower occiput. The lateral branch is motor to the posterior cervical muscles.

Both the medial and lateral branches of the posterior rami of C4–8 supply the posterior cervical muscles. In addition, the medial branches of C4 and C5 supply the overlying skin. Note that trapezius receives twigs from anterior rami of C3 and C4 in addition to its main motor innervation from the accessory nerve.

The thoracic posterior primary rami

All the thoracic posterior primary rami divide into medial and lateral branches, all of which supply the dorsal muscles. The *medial* branches of the upper six thoracic nerves reach and supply the skin immediately adjacent to the vertebral spines, whereas the *lateral* branches of the lower six thoracic nerves are cutaneous as well as motor.

The cutaneous branches descend for a distance that increases from above downwards before they supply the skin. Thus, T1 supplies an area immediately inferior to its corresponding vertebra, whereas T10 and T11 innervate the skin over the loin and T12 runs along the iliac crest, then sends twigs over the upper gluteal region.

The lumbar posterior primary rami

All the lumbar posterior primary rami divide into medial and lateral branches that supply the overlying lumbar muscles. The lateral branches of the upper three, in addition, reach the skin over the posterosuperior iliac spine and innervate the adjacent gluteal region.

The sacral and coccygeal posterior primary rami

The posterior rami of S1–4 emerge through the posterior sacral foramina; that of S5 emerges from the bifurcation of the main nerve trunk after this escapes in the gap between the cornua of the sacrum and coccyx. These nerves are small, and become successively smaller from above downwards. They all supply sacrospinalis; only the upper three divide to give lateral branches that reach the skin over the sacrum.

Co.1 is small, is undivided and, appropriately, supplies the skin over the coccyx.

The anterior primary rami

The anterior primary rami supply the arm, the leg and the front and sides of the neck, thorax and abdomen with their motor and sensory innervation. The trunk is supplied, in the main, by the segmentally placed thoracic anterior rami; the neck and limbs are served by the cervical, brachial, lumbar and sacral plexuses.

The segmental cutaneous nerve distribution of the body is shown in Fig. 125.

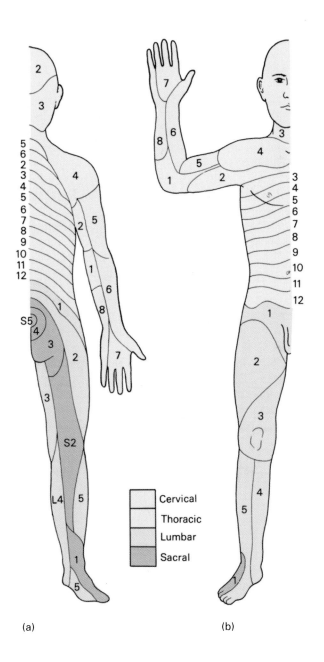

Fig. 125 The segmental cutaneous supply of the body.

(a)   (b)

The cervical plexus

The anterior rami of the upper four cervical nerves unite by a series of loops to form the cervical plexus, whose function is the supply of the skin and muscles of the neck and the innervation of the diaphragm.

Formation of the plexus (Fig. 126)

The loops are three in number, C1–2, C2–3 and C3–4, with a further loop (C4–5) often present to connect the cervical plexus with the brachial plexus. They lie on the scalenus medius and legato scapulae muscles under the cover of the sternocleidomastoid muscle.

The anterior primary ramus of C1 is entirely motor. It emerges from the vertebral canal in the groove on the posterior arch of the atlas (Fig. 92) immediately behind the superior articular facet. Here, the nerve intervenes between the posterior arch and the vertebral artery. The nerve then runs forward on the lateral side of the lateral mass, lying *medial* to the vertebral artery as this emerges from its foramen transversarium. Twigs of supply are given to rectus capitis lateralis and anterior and to longus capitis, then the nerve descends to form a loop with the ascending branch of C2 in front of the transverse process of the atlas.

The majority of the fibres in this loop run forward to join the hypoglossal nerve at the level of the atlas; through this link with XII, C1 supplies

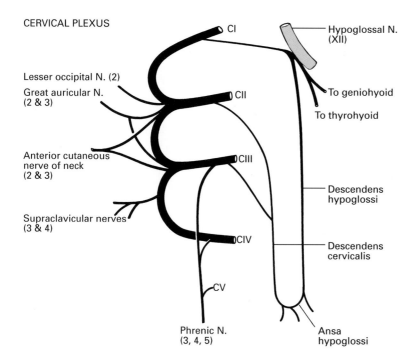

CERVICAL PLEXUS

CI

Hypoglossal N. (XII)

Lesser occipital N. (2)

Great auricular N. (2 & 3)

CII

To geniohyoid

To thyrohyoid

Anterior cutaneous nerve of neck (2 & 3)

CIII

Descendens hypoglossi

Supraclavicular nerves (3 & 4)

CIV

Descendens cervicalis

CV

Phrenic N. (3, 4, 5)

Ansa hypoglossi

Fig. 126 Plan of the cervical plexus.

the geniohyoid and thyrohyoid muscles, then runs downwards as the *descendens hypoglossi*, from which the nerve to the anterior belly of the omohyoid muscle is derived. Descendens hypoglossi joins *descendens cervicalis*, derived from C2 and C3, to form a loop termed the *ansa cervicalis*, which lies on the carotid sheath (see Fig. 211). From the ansa nerve, fibres pass to supply sternohyoid, sternothyroid and the posterior belly of omohyoid.

The anterior primary ramus of C2 emerges posteriorly to the superior articular process of the axis, then passes forwards on the lateral side of the vertebral artery. It divides into an ascending branch, which joins C1, and a descending branch, which loops to join C3.

The remaining anterior primary rami of the cervical nerves emerge from the intervertebral foramina anterior to their articular pillars and lateral to the vertebral artery. Each root receives a grey ramus communicans from the superior cervical ganglion.

Summary of branches

The branches of the cervical plexus can be divided into four groups:

1 *Communicating branches*, which pass to the hypoglossal nerve, as already described, and which also pass to the vagus and to the cervical sympathetic chain.
2 *Superficial branches*, which supply cutaneous fibres to the neck.
3 *Deep branches*, to the neck muscles.
4 The *phrenic nerve*, which is the motor nerve of the diaphragm (apart from an unimportant contribution to the crura from T11 and T12) and which also transmits proprioceptive fibres from the central part of this muscle.

Obviously, the second and fourth groups are of greatest interest and importance to the anaesthetist.

The superficial cervical plexus

The superficial cervical plexus comprises *superficial branches* (Fig. 127) that can be tabulated into ascending, transverse and descending groups:

1 *Ascending* – lesser occipital nerve (C2); great auricular nerve (C2, C3).
2 *Transverse* – anterior cutaneous nerve of neck (C2, C3).
3 *Descending* – the supraclavicular nerves (C3, C4); note that C1 has no cutaneous branch.

The *lesser occipital nerve* (C2) hooks around the spinal accessory nerve (XI), then ascends along the posterior border of the sternocleidomastoid. It pierces the deep fascia in the upper part of the posterior triangle, then splits up into three branches:

1 *Auricular* – to the upper third of the medial aspect of the external ear.
2 *Mastoid* – to the skin over the mastoid process.
3 *Occipital* – to the occipital area immediately above and behind the mastoid.

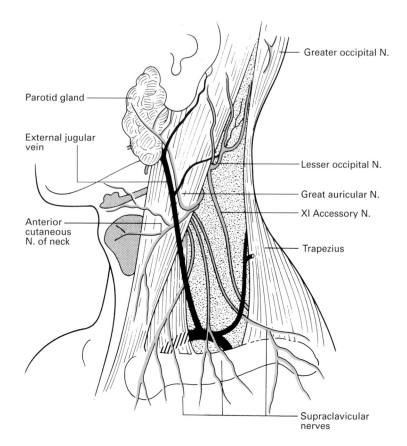

Parotid gland

External jugular vein

Anterior cutaneous N. of neck

Greater occipital N.

Lesser occipital N.

Great auricular N.

XI Accessory N.

Trapezius

Supraclavicular nerves

Fig. 127 The cutaneous branches of the cervical plexus.

The *great auricular nerve* (C2, C3) is the largest cutaneous branch of the cervical plexus. It hooks around the mid-point of the posterior border of sternocleidomastoid, then passes across it in the direction of the angle of the mandible. On this muscle it breaks up into three terminal branches:

1 *Auricular* – supplying the lower two-thirds of the medial aspect of the external ear and the lateral surface of the lobule.
2 *Mastoid* – to the skin over the mastoid process.
3 *Facial* – to the skin over the masseter and the parotid gland.

The *anterior cutaneous nerve of the neck* (C2, C3) emerges close below the great auricular nerve at the posterior border of sternocleidomastoid, then passes horizontally forwards on the muscle, deep (sometimes superficial) to the external jugular vein. At the anterior border of sternocleidomastoid, the nerve pierces the deep fascia and splits into branches to supply the skin of the whole front of the neck.

The *supraclavicular nerves* (C3, C4) (Fig. 127) arise as a common stem that emerges from behind sternocleidomastoid immediately below the other cutaneous nerves of the plexus. This stem soon splits into three branches – medial, intermediate and lateral – that pierce the deep fascia above the clavicle, cross this bone and supply the skin over the upper sternum, the upper chest wall (as far down as the 3rd rib) and the upper deltoid. On careful palpation, these nerves can be rolled over the subcutaneous anterior border of the clavicle. Note that, although the supraclavicular nerves do not form part of the brachial plexus, they are often blocked by approaches to the upper plexus, e.g. the interscalene approach. It is likely that this is the result of cranial paravertebral spread of local anaesthetic.

The deep cervical plexus

This supplies the anterior vertebral muscles – the recti capitis, longus capitis and longus cervicis, as well as giving contributions to scalenus medius (the main scalene innervation is from the roots of the branchial plexus). In addition, branches pass to levator scapulae (C3, C4) and to two muscles whose principal innervation is from the spinal accessory nerve: sternocleidomastoid (C2, C3) and trapezius (C3, C4).

CLINICAL NOTE

Superficial and deep cervical plexus blocks

Surgery in the anterior triangle of the neck, such as carotid endarterectomy, can be performed after local anaesthetic blockade of the superficial and deep cervical plexuses. The block should produce anaesthesia over the occipital region, the neck, shoulder and upper pectoral region. The superficial cervical plexus can be blocked as it emerges from behind the posterior border of the middle portion of the sternocleidomastoid muscle. The injection of 5–10 ml of local anaesthetic solution at this place is usually sufficient to provide analgesia of the skin over the anterior triangle. Blockade of the deep cervical plexus is achieved by depositing small (3–5 ml) volumes of local anaesthetic near the transverse processes of the 2nd, 3rd and 4th cervical vertebrae (C2–4). With the patient in the supine position and the head turned away from the side to be blocked, the mastoid process and the transverse process of C6 (at the level of the cricoid cartilage, the most prominent of the cervical transverse processes) are identified, and a line is drawn between them. The roots of the cervical plexus lie beneath this line. The transverse process of C2 is 1.5–2.0 cm distal to the tip of the mastoid process, that of C4 is approximately midway between the clavicle and the mastoid process, that of C3 lies midway between the transverse processes of C2 and C4 (Fig. 128). After subcutaneous infiltration of local anaesthetic, a needle is introduced perpendicular to the skin until it makes contact with the transverse process of C2, usually at a depth of 1–2 cm. After withdrawal of the needle by 1 mm, and after careful aspiration to exclude intravascular or subarachnoid needle-tip placement, 3–5 ml of local anaesthetic is injected. The injection is repeated at the C3 and C4 levels (Fig. 129). Block of the phrenic nerve is common, and this block should therefore not be performed bilaterally. A Horner's syndrome (see below) may occur as a result of a cervical sympathetic block, as may on occasion a vagus nerve block.

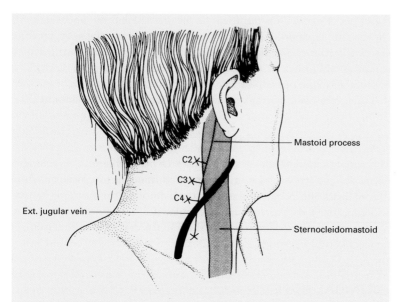

Fig. 128 The surface markings of a deep cervical plexus block. The transverse process of C2 lies 1.5–2.0 cm below the tip of the mastoid process, C4 mid-way between the clavicle and the mastoid and C3 mid-way between these two points.

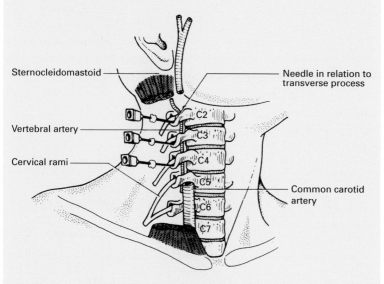

Fig. 129 The anatomy of deep cervical plexus block.

The stellate ganglion (Fig. 130)

The inferior cervical sympathetic ganglion is fused with the 1st thoracic ganglion in about 80% of subjects. This combined structure is termed the stellate ganglion, although the term is often used for the two ganglia even when they are separate. The inferior cervical ganglion itself lies in front of the anterior ramus of the 8th cervical nerve and immediately posterior to the vertebral vessels or to the upper border of the subclavian artery, if this is rather highly arched; it is at the level of the disc space between C7 and T1 vertebrae. The 1st thoracic ganglion lies against the neck of the 1st rib behind the pleura. The inferior ganglion is connected with the middle cervical ganglion both by the cervical sympathetic chain itself and by the ansa subclavia (or ansa Vieussens), which loops around the inferior margin of the subclavian artery then passes upwards to join the middle ganglion.

Preganglionic sympathetic fibres from spinal segments T1 and T2 pass upwards along the sympathetic chain to relay mainly in the superior cervical ganglion for distribution to the head and neck. Fibres from spinal segments T2–7 relay mainly in the ganglia of T1 and T2 and in the inferior and middle cervical ganglia. Grey rami stream thence to the roots of the brachial plexus and are thus distributed to the upper limb.

The close relationship of the stellate ganglion and lower cervical sympathetic chain to the brachial plexus means that diffusion of local anaesthetic solution will often produce a sympathetic blockade of the head and neck after a brachial plexus infiltration is performed. The clinical manifestation of this is a *Horner's syndrome*: the pupil is small (paralysis of the dilator pupillae), there is ptosis (paralysis of the sympathetic supply to levator

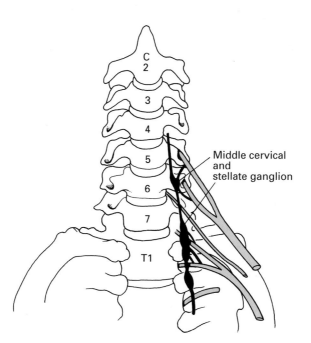

Fig. 130 The stellate ganglion shown in its relationship to the roots of the brachial plexus and the vertebrae.

palpebrae) and there is unilateral vasodilatation and anhidrosis owing to blockage of the sympathetic outflow to the skin of the face; the vasodilatation also causes unilateral nasal blockage. There is doubt whether the enophthalmos described in this syndrome actually occurs.

In performing a sympathetic denervation of the upper limb (upper dorsal sympathectomy), the surgeon divides the sympathetic chain immediately below the T3 ganglion, then dissects up the chain, dividing all its connections, but carefully preserving the stellate ganglion with its white ramus from T1. In this way, the sympathetic outflow to the upper limb (T2–7) is cut off by preganglionic section, but the main supply to the head and neck via T1 is preserved; by this manoeuvre the rather unsightly Horner's syndrome is avoided. The thoracoscopic approach gives an excellent view of the upper thoracic sympathetic chain and both sides can be operated upon at the same session.

Some sympathetic fibres to the upper limb may leave the chain below the ganglion of T1 and run directly to the brachial plexus. One fairly constant strand from the ganglion of T2 to the 1st thoracic nerve is termed the *nerve of Kuntz*; it must be sought and divided if the sympathectomy is to be complete.

CLINICAL NOTE

Stellate ganglion block

Indications for stellate ganglion block include pain syndromes such as refractory angina pectoris, complex regional pain syndromes in the arm, phantom limb pain, post-herpetic neuralgia and shoulder/hand syndrome. Vascular insufficiency in the arm, acute or chronic, may benefit from stellate ganglion block. The causes include frostbite, Raynaud's syndrome, trauma, emboli, scleroderma and obliterative vascular disease. In obese patients, people with very short necks and those in whom the anatomy is altered, image intensification should be used. It is best to have an assistant to aspirate and inject on command, leaving the operator to fix the needle position. The patient lies in the dorsal recumbent position with the neck in extension. The mouth is opened slightly to relax the neck muscles. The patient should be told not to talk or move their neck during the procedure. The transverse process of the 6th cervical vertebra (Chassaignac's tubercle) is palpated at the level of the cricoid cartilage. The carotid artery and sternocleidomastoid muscle may need to be retracted laterally in order to palpate the tubercle (Fig. 131). After raising a skin wheal of local anaesthetic, a needle is passed perpendicularly backwards until it impinges onto the transverse process of the 6th cervical vertebra. The exact depth varies with the build of the patient but may be up to 3 cm from the skin surface. After contacting bone, the needle is withdrawn 1–2 mm. The bevel of the needle should be pointing in a caudal direction. Confirmation of placement may be made using fluoroscopy, ultrasound or using computed tomography imaging. Injection of contrast should confirm extravascular placement, showing

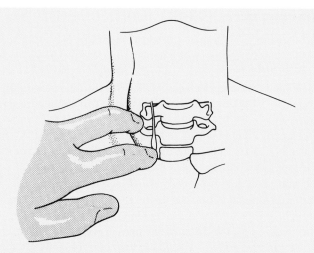

Fig. 131 The anatomy of stellate ganglion block. The fingers retract the sternocleidomastoid and the common carotid artery.

the dye spreading cephalic and caudal in a characteristic fashion. If intravenous or intra-arterial the dye will disappear suddenly, and if the injection is intramuscular no significant spread will occur. Aspiration should take place further to exclude subarachnoid or intravascular placement. A test dose of 0.5 ml of solution is injected; however, the utility of this has been questioned as seizures may occur even with such volumes if the needle is in a vertebral artery. The patient should communicate by blinking or moving a finger that there are no adverse effects. A 2–3 ml injection of an epinephrine-containing local anaesthetic will confirm intravenous placement. The remainder of the injectate (up to 10–15 ml) should be given in increments with repeated aspiration and with continual visual communication with the patient. Confirmation of the adequate placement of the local anaesthetic is by the development of a Horner's syndrome (Fig. 132).

Fig. 132 Diagram of the ptosis and meiosis produced by a left stellate ganglion block.

The phrenic nerve (C3–5)

The *phrenic nerve* is, of course, the most important branch of the cervical plexus. It provides the motor innervation of the diaphragm (apart from a clinically insignificant contribution to the crura from T11 and T12), and transmits proprioceptive sensory fibres from the central part of the diaphragm. In addition, filaments are supplied to the pleura and pericardium.

The principal component of the nerve is derived from the anterior primary ramus of C4, but contributions are also provided from C3 and C5. The three roots of the nerve join at the lateral border of scalenus anterior and then the fully constituted nerve runs downwards and medially across the anterior face of the muscle, covered by, and showing through, the prevertebral fascia. On scalenus anterior, the phrenic nerve is overlapped by the internal jugular vein and the sternocleidomastoid muscle, and is crossed by the inferior belly of the omohyoid and by the transverse cervical and transverse scapular vessels. On the left side, in addition, the nerve is crossed by the thoracic duct (Fig. 133).

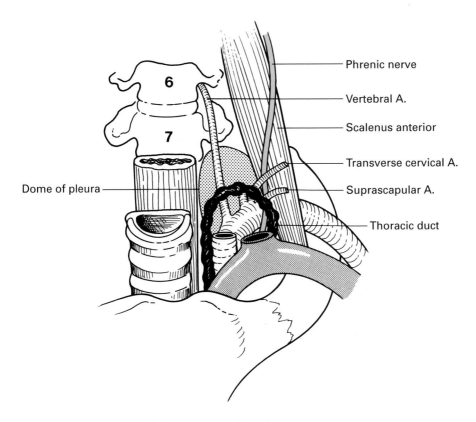

Fig. 133 Dissection of the left phrenic nerve in the neck.

The nerve then passes over the first part of the subclavian artery, behind the subclavian vein, to enter the thorax, where it crosses the internal thoracic artery posteriorly from the lateral to the medial side. This artery provides a pericardiacophrenic branch that accompanies the nerve on its intrathoracic course.

Within the thorax, the relations of the nerve differ on each side. On the right (Fig. 54), the nerve hugs the great venous pathway, descending on the lateral sides, successively, of the right brachiocephalic vein, the superior vena cava, the right atrium (separated by pericardium) and the intrathoracic portion of the inferior cava, covered throughout laterally by the mediastinal pleura. On the left (Fig. 55), the nerve has a longer and more oblique course. It passes down between the left subclavian and left common carotid arteries, crosses the arch of the aorta (passing here in front of the vagus nerve), descends anterior to the root of the lung and then along the pericardium covering the left ventricle. Laterally lies the mediastinal pleura.

On the right, the nerve pierces the central tendon of the diaphragm immediately lateral to the opening for the inferior vena cava; some nerve fibres may actually accompany the vein through this orifice. The left nerve penetrates the diaphragm 1 cm lateral to the attachment of the fibrous pericardium (Fig. 62). On both sides, the nerve fibres then supply the muscle on its abdominal aspect.

Occasionally, the contribution from C5 to the phrenic nerve may come as an accessory phrenic nerve, either directly from the root of C5 across scalenus anterior or from the nerve to subclavius. In the latter case, the filament crosses anteriorly (occasionally posteriorly) to the subclavian vein to join the main phrenic trunk behind the 1st costal cartilage.

The brachial plexus

The brachial plexus provides the motor innervation and nearly all the sensory supply of the upper limb.

Formation of the plexus (Fig. 134)

The plexus is formed by the anterior primary rami of C5–8, together with the bulk of T1. In addition, there is frequently a contribution above from C4 and below from T2. Occasionally, the plexus is mainly derived from C4–8 (pre-fixed plexus) or from C6–T2 (post-fixed plexus), variations that are usually associated with the presence of a cervical rib or of an anomalous 1st rib, respectively (see page 297).

The five *roots* of the plexus emerge from the intervertebral foramina. Each of those from C5, C6 and C7 passes behind the foramen transversarium of its respective cervical vertebra with its contained vertebral vessels, then lies in the gutter between the anterior and posterior tubercles of the corresponding transverse process. All five roots then become sandwiched

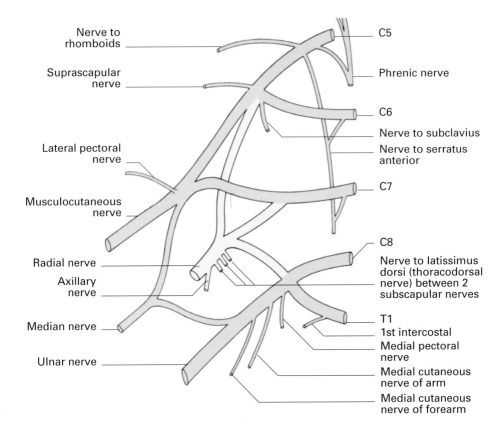

Fig. 134 Plan of the brachial plexus.

between scalenus anterior and medius (Fig. 135). Here, the roots of C5 and C6 unite into the *upper trunk*, the root of C7 continues as the *middle trunk* and those of C8 and T1 link into the *lower trunk*. As the roots of the brachial plexus emerge in the groove between the anterior and posterior tubercles of the transverse processes of the cervical vertebrae, they lie in a fibro-fatty space between two sheaths of fibrous tissue. The posterior part of the sheath arises from the posterior tubercles and covers the front of scalenus medius; the anterior part arises from the anterior tubercles and covers the posterior aspect of scalenus anterior. Laterally, the sheath extends as a covering around the brachial plexus as this emerges into the axilla (Fig. 136). The significance of this space to the anaesthetist is that it forms a sheath around the brachial plexus into which local anaesthetic can be injected to produce a brachial plexus block. The interscalene brachial plexus block technique therefore targets the *trunks* of the brachial plexus as they pass between the scalene muscles. The sheath is, conceptually at least, continuous as far as the axilla and forms the theoretical basis of single-shot brachial plexus blocks, in that the local anaesthetic should be contained by the sheath, thereby acting on all the nerves within the sheath at that level.

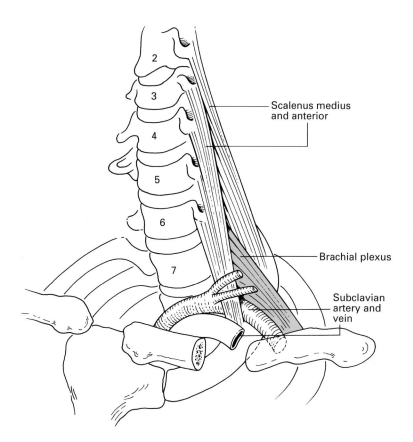

Fig. 135 The relations of the brachial plexus in the root of the neck.

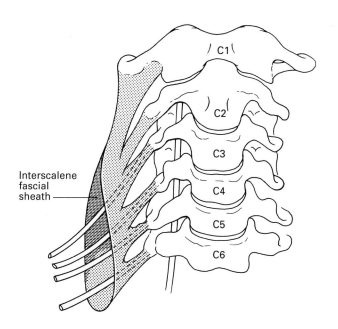

Fig. 136 The interscalene sheath. The emerging roots are enclosed in a fascial sheath.

The three trunks emerge from between the scalene muscles and pass in a closely grouped cluster downwards and laterally across the base of the posterior triangle and then across the 1st rib. As the plexus passes from between the scalene muscles to the 1st rib, it is amenable to a variety of nerve block approaches that include the supraclavicular, intersternoclei-domastoid and subclavian perivascular techniques. At the lateral border of the 1st rib, behind the clavicle, each trunk divides into an anterior and posterior *division*. The six divisions stream into the axilla and there join up into three *cords*, lateral, medial and posterior, named after the relationship they bear to the axillary artery. Between the 1st rib and axilla, a variety of brachial plexus blocks have been described: infraclavicular, vertical infra-clavicular and coracoid. The cords are composed as follows:

1 The *lateral cord* is formed by the union of the anterior divisions of the upper and middle trunks.
2 The *medial cord* represents the continuation of the anterior division of the lower trunk.
3 The *posterior cord* comprises all three posterior divisions.

The lateral and medial cords give off, respectively, the lateral and medial heads of the median nerve at the lateral border of pectoralis minor, beyond which the cords end in their terminal branches – the lateral cord becomes the musculocutaneous nerve, the medial cord the ulnar nerve and the pos-terior cord the radial and axillary nerves.

The composition of the brachial plexus can be summarized thus:

1 Five roots (between the scalene muscles) – the anterior primary rami of C5–8 and T1.
2 Three trunks (in the posterior triangle):
   a *upper*, C5 and C6;
   b *middle*, C7 alone;
   c *lower*, C8 and T1.
3 Six divisions (behind the clavicle) – each trunk divides into an anterior and posterior division.
4 Three cords (within the axilla):
   a *lateral*, the fused anterior divisions of the upper and middle trunks (C5–7);
   b *medial*, the anterior division of the lower trunk (C8, T1);
   c *posterior*, formed by the union of the posterior divisions of all three trunks (C5–T1).

The relations of the brachial plexus (Fig. 135)

Roots

Between the scalene muscles. The *roots* of the plexus lie above the second part of the subclavian artery.

Trunks

In the posterior triangle, the *trunks* of the plexus, invested in a sheath of prevertebral fascia, are superficially placed, being covered only by skin, platysma and deep fascia. However, they are crossed by a number of structures – the inferior belly of omohyoid, the external jugular vein, the transverse cervical artery and the supraclavicular nerves. Within the posterior triangle, in the thin subject, the trunks are easily rolled under the palpating fingers. The upper and middle trunks lie above the subclavian artery as they stream across the 1st rib, but the lower trunk lies behind the artery and may groove the rib immediately posterior to the subclavian groove (see page 295 and Fig. 217).

Divisions

At the lateral border of the 1st rib, the trunks bifurcate into *divisions* that are situated behind the clavicle, the subclavius muscle and the suprascapular vessels (which lie immediately posterior to the clavicle) and then descend into the axilla.

Cords

The *cords* are formed at the apex of the axilla and become grouped around the axillary artery; at first, the medial cord lies behind the artery with the posterior and lateral cords lateral to this vessel, but behind pectoralis minor the cords take up their relations to the artery as signified by their names (Fig. 137).

The branches of the brachial plexus

It is convenient for reference to enumerate here the connections and branches of the plexus.

1 The *roots* receive:

  a grey rami from the cervical sympathetic chain;

  b C5 and C6 from the middle cervical ganglion;

  c C7 and C8 from the inferior cervical ganglion;

  d T1 from the ganglion of T1;

  and give branches:

  a to longus cervicis (C5–8);

  b to the scalene muscles (C5–8);

  c nerve to rhomboids (C5);

  d nerve to serratus anterior (C5–7);

  e contribution to the phrenic nerve (C5).

2 The *trunks* give:

  a nerve to subclavius (C5, C6);

  b suprascapular nerve (C5, C6).

3 The *cords* give:

  a *lateral cord*

      lateral pectoral nerve (C5–7);

      musculocutaneous nerve (C5–7);

      lateral head of median nerve (C6, C7);

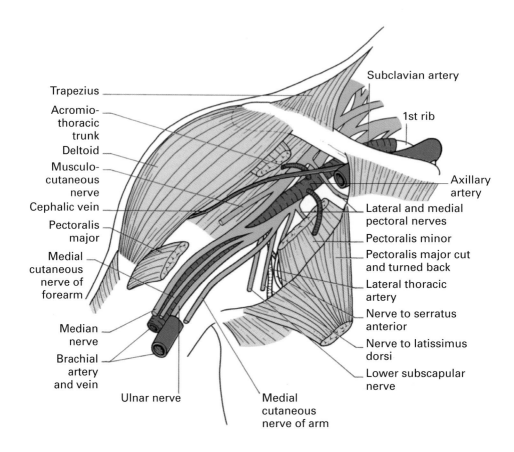

Subclavian artery

Trapezius

Acromio-
thoracic
trunk

1st rib

Deltoid

Musculo-
cutaneous
nerve

Axillary
artery

Cephalic vein

Lateral and medial
pectoral nerves

Pectoralis
major

Pectoralis minor

Pectoralis major cut
and turned back

Medial
cutaneous
nerve of
forearm

Lateral thoracic
artery

Nerve to serratus
anterior

Median
nerve

Nerve to latissimus
dorsi

Brachial
artery
and vein

Lower subscapular
nerve

Ulnar nerve

Medial
cutaneous
nerve of arm

Fig. 137 Dissection of the upper arm to show the course of the major nerves.

b *medial cord*

    medial pectoral nerve (C8, T1);
    medial cutaneous nerve of arm (C8, T1);
    medial cutaneous nerve of forearm (C8, T1);
    medial head of median nerve (C8, T1);
    ulnar nerve (C7–8, T1);

c *posterior cord*

    upper subscapular nerve (C5, C6);
    nerve to latissimus dorsi (thoracodorsal nerve) (C6–8);
    lower subscapular nerve (C5, C6);
    axillary nerve (C5, C6);
    radial nerve (C5–8, T1).

    The branches from the roots and trunks of the plexus arise in the neck –
the supraclavicular branches – but the major distribution of the plexus is
derived from its infraclavicular branches.

CLINICAL NOTE

Brachial plexus blocks

Brachial plexus blocks are techniques that aim to place local anaesthetic solutions around the nerves of the brachial plexus. These extend from approaches to the upper trunks of the plexus as they pass from the cervical spine down between the anterior and middle scalene muscles (the *interscalene* approach), through periclavicular approaches that aim to block the whole plexus as it passes behind the clavicle, to infraclavicular and axillary approaches that aim to target the terminal nerves supplying the forearm and hand. The choice of technique rests upon many factors, not least of which are the target area for blockade and the risk of pneumothorax that inevitably accompanies periclavicular blocks. The suprascapular nerve provides much of the nerve supply to the shoulder joint. It is a relatively early branch of the brachial plexus (upper trunk), and, as a result, successful blocks for shoulder surgery must be aimed at the proximal portion of the brachial plexus. Interscalene brachial plexus anaesthesia is currently the most popular approach to this target area and is routinely used in the perioperative management of shoulder surgery.

Interscalene brachial plexus block (Fig. 138)

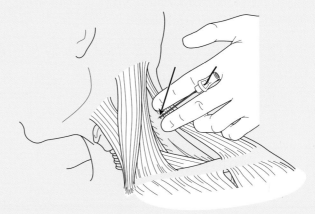

Fig. 138 Interscalene brachial plexus block. The needle position demonstrates the traditional approach; the arrow indicates a more caudal angulation that is often recommended.

Contemporary practice involves the usage of ultrasound to provide real-time anatomical guidance with or without a peripheral nerve stimulator and nerve-stimulating needles to confirm proximity to the nerves. A further development is the use of echogenic needles to enhance the ultrasound view of the needle. The use of ultrasound is associated with a decreased onset time and a reduction in the number of needle passes. In this technique, the brachial plexus is blocked at the level of the sixth cervical vertebra. The interscalene groove that lies between the anterior and medial scalene muscles is identified at this level. The roots and proximal trunks of the brachial plexus are sandwiched between the anterior and medial scalene muscle. The needle is passed at an angle that is classically described as being 'at right angles to the skin in all planes'. With ultrasound the needle may be passed in plane, in which the needle should be visible along the entire length, or out of plane, in which the passage of the needle is deduced by tissue movement. Following ultrasonic localization a nerve stimulator can be used to confirm nerve localization by evoking contractions in the deltoid or biceps brachialis muscles. Injection of up to 30 ml of local anaesthetic solution can then be performed with, more recently, the addition of adjuncts such as clonidine or dexamethasone to attempt to prolong the duration of the block. Although cervical plexus block, cervical sympathetic and phrenic nerve block are so common as to be termed 'accompaniments' to the block rather than complications of the block, there are a number of serious complications that are ordained by the proximity of the brachial plexus to other anatomical structures. Therefore, intravascular (vertebral artery and external jugular vein), epidural and subarachnoid injections are possible, with potentially serious consequences. Access of the injecting needle to the neuraxis is a particular concern. Interscalene blocks will give excellent analgesia of the shoulder and outer arm and distal clavicle. They are less effective for hand surgery, as the lower trunks of the brachial plexus are often missed by the local anaesthetic. Interscalene catheter techniques have also been described.

Subclavian perivascular brachial plexus block (Fig. 139)

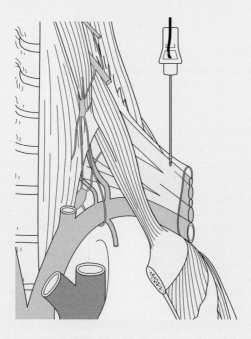

Fig. 139 Subclavian perivascular block.

This technique targets the brachial plexus as it passes over the 1st rib and can be used to provide analgesia of the whole arm, although it is occasionally deficient in the territory ultimately supplied by the ulnar nerve. Any periclavicular approach is associated with the risk of puncture of the pleura and consequently of a pneumothorax. This technique aims to minimize the risk of a pneumothorax by using the 1st rib as a 'backstop'. The interscalene groove is identified at the level of the 6th cervical vertebra and is followed down the neck to the point at which the scalene muscles insert onto the 1st rib. The injection is made just posterior to the pulsation of the subclavian artery if this is felt. The needle angle is, most commonly, in plane under ultrasonic guidance along the axis of the clavicle, and nerve stimulation produces contractions of the muscles of the forearm or hand. Again, up to 30 ml of local anaesthetic is injected. Pneumothorax remains a complication of this technique, but the incidence of this complication should be low in experienced hands. Intravascular injection is another noted complication because of the proximity of the subclavian vessels.

Axillary brachial plexus block (Fig. 140)

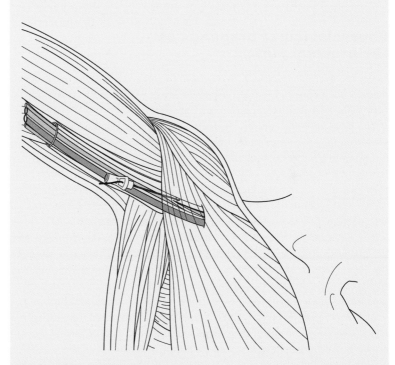

Fig. 140 Axillary brachial plexus block.

This technique targets three of the four major terminal nerves of the brachial plexus (ulnar, radial and median nerves) as they pass with the axillary artery and vein along the humerus high in the axilla. The nerves supplying the shoulder and upper arm have left the plexus at this point and, as a result, this block is most appropriate for anaesthesia of the elbow, forearm and hand. The musculocutaneous nerve also usually leaves the plexus above this point and is therefore often blocked separately. The block can be performed with a 'single injection' technique that relies upon the concept of a fascial sheath that surrounds the nerves and vessels. Alternatively, the three nerves can be identified and blocked individually. With the patient's arm abducted so as to be at 90° to the trunk, the pulsation of the axillary artery is identified as high as possible in the axilla. A needle is inserted to one side of the artery and paraesthesiae are sought in the hand or, if a nerve stimulator is used, contractions of muscles supplied by one of the three nerves are evoked. A single injection of 30–60 ml of local anaesthetic is performed or, if a multiple injection technique is being used, the nerves are identified individually and 10 ml of local anaesthetic solution is injected around each nerve.

There are other approaches to the brachial plexus, but space precludes their individual consideration. The reader is referred to one of the many textbooks on regional anaesthesia.

Supraclavicular branches of the brachial plexus

The *nerve to the rhomboids* (C5) arises from the root of C5. It pierces scalenus medius, then crosses the deep aspect of levator scapulae to reach and to supply the rhomboids.

The *nerve to serratus anterior* (C5–7), or the nerve of Bell, receives its origins from the anterior primary rami of C5, C6 and C7, although the contribution from C7 is inconstant; those from C5 and C6 pierce scalenus medius, that from C7 passes across the anterior face of this muscle. The three roots then join; the common trunk thus formed passes over the 1st rib then continues over serratus anterior, lying rather posteriorly placed on the medial wall of the axilla, to supply each digitation of this muscle.

The *suprascapular nerve* (C5, C6) originates from the upper trunk, crosses the posterior triangle, then passes deep to trapezius. The nerve traverses the suprascapular notch (which is arched over by the suprascapular ligament) and descends deep to the supraspinatus and infraspinatus, both of which muscles it supplies. It also supplies sensation to the shoulder joint.

The *nerve to subclavius* (C5, C6) is a small nerve that arises from the upper trunk, descends in front of the plexus and the third part of the subclavian artery, behind the subclavian vein, to supply subclavius as this lies in its groove below the clavicle. The important point about this nerve is its occasional contribution to the phrenic nerve (see page 165).

Infraclavicular branches of the brachial plexus

These are conveniently considered according to their cords of origin.

Branches of the lateral cord (Fig. 137)

The *lateral pectoral nerve* (C5–7) crosses the axillary vessels, pierces the clavipectoral fascia and enters pectoralis major, which it supplies. The clavipectoral fascia is the tough membrane that stretches from pectoralis minor below, which it encloses, to the clavicle above, ensheathing subclavius. In addition, the fascia is pierced by the cephalic vein as this passes to its termination in the axillary vein, by the acromiothoracic trunk of the axillary artery, and by lymphatics.

A twig from the lateral pectoral nerve crosses in front of the axillary vessels to join the medial pectoral nerve through which the lateral nerve contributes a supply to pectoralis minor.

The *musculocutaneous nerve* (C5–7) is the continuation of the lateral cord after this has given off the lateral head of the median nerve at the lower border of pectoralis minor. Because of its derivation from the lateral cord, the nerve naturally lies lateral to the axillary artery. It first supplies and then pierces coracobrachialis, then descends downwards and laterally between biceps and brachialis, supplying both of these muscles. (Brachialis, in addition, receives a nerve supply from the radial nerve; see page 187.) The nerve emerges between the biceps tendon and brachioradialis, pierces the deep fascia of the antecubital fossa and continues downwards as the *lateral cutaneous nerve of the forearm*. This divides into an anterior branch, which passes anterolaterally over the forearm to the base of the thenar eminence, and a posterior branch, which descends over the posterolateral aspect of the forearm to the wrist.

The musculocutaneous nerve in addition sends a twig to the elbow joint, thus acting in accordance with Hilton's law, which states that 'the same trunks of nerves whose branches supply the groups of muscles moving a joint, furnish also a distribution of nerves to the skin over the insertions of the same muscles; and the interior of the joint receives its nerves from the same source'.

Branches of the medial cord (Fig. 137)

The *medial pectoral nerve* (C8, T1) arises from the medial cord at the axillary apex. Passing forwards between the axillary artery and vein, it receives a twig from the lateral pectoral nerve, then pierces and supplies pectoralis minor; some fibres pass on to supply the overlying pectoralis major.

The *medial cutaneous nerve of the arm* (C8, T1) is the smallest branch of the brachial plexus. Originating between the axillary artery and vein, it crosses either in front of or behind the vein to run along its medial aspect – the only nerve to do so. The nerve receives a communicating branch from the intercostobrachial nerve (the lateral cutaneous branch of the 2nd intercostal nerve), then pierces the deep fascia at the mid-point of the arm to supply the skin over the distal half of the medial side of the arm as far as the elbow.

The *medial cutaneous nerve of the forearm* (C8, T1) descends first between the axillary artery and vein and then on the medial side of the brachial

artery. At the mid-point of the arm it pierces the deep fascia and divides
into two branches. The anterior branch crosses usually in front of, although
sometimes behind, the median cubital vein in the antecubital fossa (see
Fig. 218) then descends to supply the skin over the anteromedial side of
the forearm to the wrist. The posterior branch descends on the medial side
of the basilic vein, then passes onto the back of the forearm to innervate its
posteromedial aspect as far as the wrist.

The *ulnar nerve* (C7, C8, T1), shown in Figs 137 and 141, is the continuation of the medial cord after this has given off the medial head of the

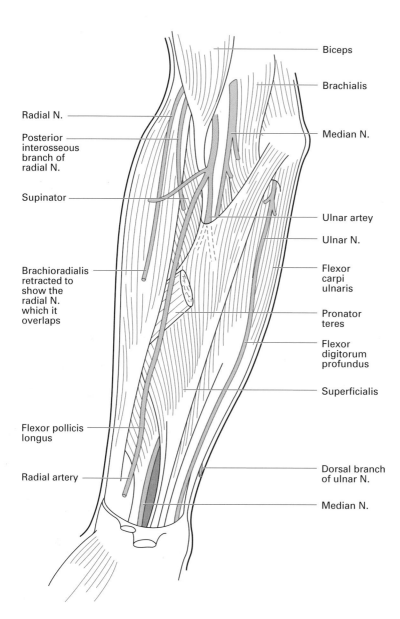

Fig. 141 Dissection of
the front of the right
forearm to show the
course and distribution
of the principal nerves.

median nerve. The ulnar nerve is usually composed of fibres from C7, C8 and T1, but in some 15% of cases there is no C7 contribution.

The nerve runs first between the axillary artery and vein and then along the medial aspect of the brachial artery as far as the insertion of coraco-brachialis into the humerus, at approximately the mid-point of the arm. Here, it passes backwards to pierce the medial intermuscular septum and continues its descent on the anterior face of the medial head of the triceps. The nerve passes behind the medial epicondyle of the humerus, where it can be rolled against the bone (Fig. 142), then dives between the humeral and ulnar heads of flexor carpi ulnaris, lying here on the capsule of the medial aspect of the elbow joint.

Together with the median nerve, the ulnar nerve is vulnerable to injury in forcible abduction of the arm. This may occur when the arm is placed more than 90° from the trunk during axillary dissection or operations upon the breast. It is wise, therefore, never to allow the arm of the anaesthetized

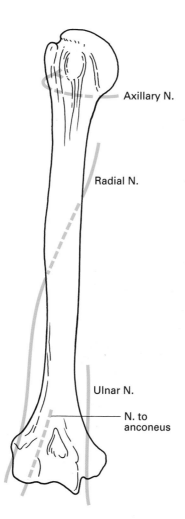

Axillary N.

Radial N.

Ulnar N.

N. to
anconeus

Fig. 142 The nerves
related to the humerus.

patient to be placed at more than 90° to the trunk and it is a useful pre-
caution to have the hand pronated when the arm is abducted in order to
reduce the stretching effect on the nerve trunks. The close relationship of
the nerve to the elbow joint and its course around the medial epicondyle
are of considerable practical importance. In this position it is particularly
susceptible to pressure injuries from the edge of the operating table or from
arm-retaining straps or clips. Because of its proximity to easily identified
bony landmarks, it is a convenient place to perform an ulnar nerve block
or to place needle electrodes to stimulate the nerve electrically in order to
detect the presence or absence of block at the ulnar neuromuscular junction. However, injection of local anaesthetic into or around the ulnar nerve
directly as it passes around the medial epicondyle is thought to be associated with an increased incidence of neuritis, and, for this reason, most
anaesthetists aim to perform ulnar nerve block at the elbow some 1–3 cm
above the medial epicondyle.

In its course in the posterior compartment of the upper arm and behind
the elbow, the nerve is accompanied by a branch of the brachial artery,
appropriately named the ulnar collateral artery, which takes part in the rich
vascular anastomosis around the elbow. In the forearm, the ulnar nerve
descends on flexor digitorum profundus. In the upper forearm, the nerve
is covered by flexor carpi ulnaris but, as the latter becomes tendinous, it
lies on the lateral side of the nerve, which is then covered only by skin
and deep fascia. The nerve then crosses the flexor retinaculum immediately lateral to the pisiform bone, covered here by some superficial fibres
of the retinaculum. This forms a tunnel, Guyon's canal, where the nerve
may occasionally become entrapped. At this point, the close relationship
of the ulnar nerve to an easily palpable bony landmark, the pisiform bone,
renders it available for nerve block (as part of wrist block) and also for
electrical stimulation in the detection of suspected neuromuscular block. It
crosses the hook of the hamate bone (against which it can be compressed to
produce a dull 'nerve' pain) and then divides into its superficial and deep
terminal branches.

The ulnar artery approaches the nerve from the lateral side in the forearm, meets it about a hand's breadth below the elbow and then accompanies it along its lateral side through the forearm and wrist.

The ulnar nerve supplies the following:

1 *Muscular branches* – to flexor carpi ulnaris, the medial half of flexor digitorum profundus, and all the intrinsic muscles of the hand apart from
the lateral two lumbricals and the three muscles of the thenar eminence.
2 *Cutaneous branches* – to the dorsal and palmar aspects of the medial side
of the hand and of the medial $1\frac{1}{2}$ fingers.
3 *Articular branches* – to elbow and wrist.

(Note that no branches arise in the upper arm. The muscular branches
to flexor carpi ulnaris and the medial half of flexor digitorum profundus
are given off immediately below the elbow – the lateral half of profundus
is supplied from the anterior interosseous branch of the median nerve.)

Two cutaneous branches are given off in the forearm:

1 The *palmar cutaneous nerve* arises in the mid-forearm, descends along
the anterior aspect of the ulnar artery, pierces the deep fascia above

the wrist, crosses the flexor retinaculum and supplies the skin over the hypothenar eminence.

2 The *dorsal branch* arises 5.0–7.5 cm above the crease of the wrist, passes backwards under cover of the tendon of flexor carpi ulnaris, perforates the deep fascia above the wrist and supplies the ulnar border of the dorsum of the hand. It then breaks up into dorsal digital nerves – one to the dorsum of the ulnar side of the 5th finger, a second that bifurcates at the metacarpal head level to supply the backs of the adjacent sides of the 4th and 5th fingers, and an occasional third branch that usurps radial nerve territory to supply the adjacent sides of the 3rd and 4th fingers.

The dorsal digital nerves fail to reach the skin over the posterior aspects of the distal phalanges (and therefore the nail beds); these are supplied by collateral branches from the palmar digital nerves (ulnar to the 5th and ulnar side of the 4th, and median to the radial side of the 4th finger).

In the hand (Fig. 143), for practical purposes, the superficial terminal branch can be considered sensory (apart from a motor twig to palmaris brevis) and the deep branch motor (apart from a sensory twig to the wrist joint).

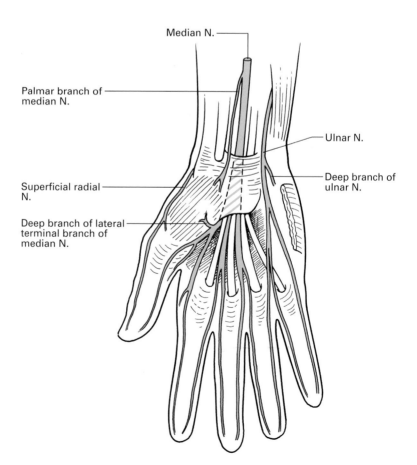

Median N.

Palmar branch of
median N.

Ulnar N.

Deep branch of
ulnar N.

Superficial radial
N.

Deep branch of lateral
terminal branch of
median N.

Fig. 143 The
distribution of the
nerves in the front of
the hand.

The *superficial terminal branch*, after supplying the overlying palmaris brevis, splits into two palmar digital nerves. These pass distally under cover of the tough palmar aponeurosis and are deep to the corresponding arterial digital branches of the superficial palmar arch.

The medial digital branch supplies the ulnar border of the 5th finger; the lateral branch bifurcates to supply the adjacent sides of the 5th and 4th fingers. Dorsal collateral branches supply the dorsal aspects of the terminal phalanges of the 5th and the ulnar half of the 4th finger.

Each digital artery passes behind its corresponding digital nerve along the fingers, where the nerve lies alongside the flexor sheath in a plane immediately anterior to the phalanx. Incisions along the side of the finger carried straight down onto the bone thus miss both nerve and artery; conversely, a needle inserted for a digital nerve block should be placed just anterior to the anterolateral margin of the phalanx (Fig. 144).

The *deep terminal branch* is accompanied by the deep branch of the ulnar artery and plunges into the hypothenar eminence between flexor digiti minimi and abductor digiti minimi. It then pierces opponens digiti minimi and traverses the palm in company with the deep palmar arch. The surface marking is simple – the nerve continues the line of the fully extended thumb across the palm. In its course, the nerve supplies the three muscles of the hypothenar eminence, the ulnar two lumbricals, the interossei and, finally, the adductor pollicis.

Because of the increasing use of the adductor pollicis twitch response to ulnar nerve stimulation in monitoring neuromuscular block, it is important to note the surface marking of this nerve at the wrist. Here, it lies at the level of the skin crease between the palm and the wrist, immediately on the radial side of the easily palpable pisiform bone. Stimulation of the nerve at this site causes contraction of the hypothenar muscles and adductor pollicis.

Variations in the distribution of the ulnar nerve are considered on page 188.

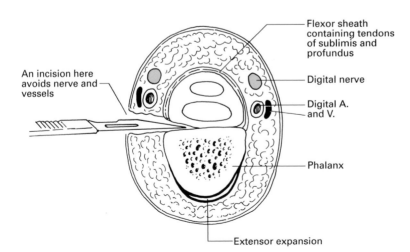

An incision here avoids nerve and vessels

Flexor sheath containing tendons of sublimis and profundus

Digital nerve

Digital A. and V.

Phalanx

Extensor expansion

Fig. 144 Transverse section of a finger showing the relationships of the digital nerve and vessels to the phalanx.

The *median nerve* (C5–8, T1), shown in Figs 137, 141 and 218, carries fibres from all the roots of the brachial plexus; C5, C6 and C7 from the lateral head, which is derived from the lateral cord of the plexus, and C8 and TI from the medial head, derived from the medial cord. The two origins of the nerve unite in front of the third part of the axillary artery. The nerve descends through the arm, first on the lateral side of the brachial artery, then on its medial side, crossing the artery at the mid-point of the upper arm at the insertion of coracobrachialis. Usually, the nerve passes across the front of the artery, but occasionally crosses behind it in the proportion of 7:1. In the rare cases of a high bifurcation of the brachial artery in the arm, the median nerve passes between the high origins of the radial and ulnar arteries.

This is a convenient place to bring together the numerous happenings at the mid-point of the upper arm:

1 Coracobrachialis is inserted into the medial side of the shaft of the humerus.
2 The lowermost fibres of deltoid reach their insertion on the lateral side of the humerus.
3 The nutrient artery from profunda brachii enters the bone.
4 The basilic vein on its course up the arm pierces the deep fascia.
5 The medial cutaneous nerve of the forearm pierces the deep fascia on its course down the arm.
6 The ulnar nerve, accompanied by the ulnar collateral branch of the brachial artery, passes backwards from the anterior to the posterior compartment of the arm through the medial intermuscular septum.

The median nerve crosses the front of the elbow lying on the brachialis; here, the nerve passes deep to the bicipital aponeurosis and the median cubital vein (page 301 and Fig. 218). It then dives between the two heads of pronator teres; the deep ulnar head of this muscle separates the nerve from the ulnar artery. This artery must cross deep to the median nerve from the lateral to the medial side in its path towards the ulnar nerve, which it is to accompany in the lower two-thirds of the forearm. The median nerve descends through the forearm (see Figs 141, 146) between the muscle bellies of flexor digitorum superficialis and profundus; it adheres closely to the posterior aspect of the former, which enables ready identification of the nerve by the surgeon. It is accompanied by the median branch of the anterior interosseous branch of the ulnar artery.

As the muscle belly of flexor digitorum superficialis narrows down to its tendons, the nerve comes to lie superficially, covered only by skin and deep fascia, with the tendon of flexor carpi radialis lying laterally and those of flexor digitorum superficialis and palmaris longus medially, the last overlapping the nerve. Here, at the skin crease of the wrist, the nerve lives up to its name – it lies exactly in the median plane (Fig. 145); a line drawn proximally from the outstretched middle finger will pass along it.

The median nerve passes into the hand through the carpal tunnel close to the deep aspect of the flexor retinaculum. In this confined space it may readily become compressed (*carpal tunnel syndrome*). Immediately beyond

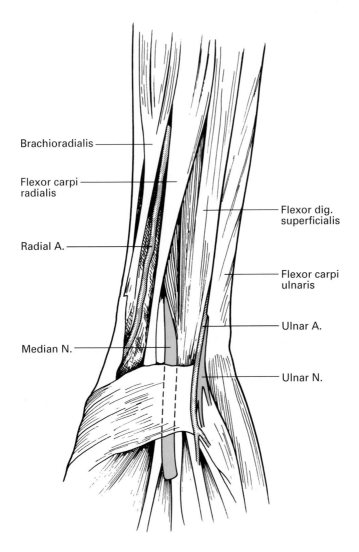

Fig. 145 The relations of the median nerve at the wrist; palmaris longus is absent in this specimen.

the retinaculum, the nerve splits up into its terminal medial and lateral branches.

The median nerve is most easily anaesthetized by infiltrating local anaesthetic solution just proximal to the flexor retinaculum and lateral to palmaris longus tendon. The needle is in the exact midline of the prominent transverse distal wrist skin crease.

The median nerve supplies the following:

1 *Muscular branches* – to pronator teres, flexor carpi radialis, palmaris longus, flexor digitorum superficialis, the three muscles of the thenar eminence and the lateral two lumbricals. Its anterior interosseous branch innervates flexor pollicis longus, the lateral half of flexor digitorum profundus and pronator quadratus.

2 *Cutaneous branches* – to the thenar eminence of the palm, the anterior aspects of the radial three digits and the skin over the dorsal aspect of their distal phalanges.

3 *Articular branches* – to the elbow and wrist.

(The nerve to pronator teres usually arises above the elbow joint; it is the only branch of the median nerve above the forearm. Branches to flexor carpi radialis, palmaris longus and flexor digitorum superficialis are given off immediately distal to the elbow joint.)

The *anterior interosseous nerve* (Fig. 146) originates where the median nerve emerges between the heads of pronator teres. It descends in company with the anterior interosseous branch of the ulnar artery along the anterior face of the interosseous membrane between radius and ulna. It supplies flexor pollicis longus, the radial half of flexor digitorum profundus (the ulnar nerve supplying the ulnar half; see page 178) and pronator quadratus; it terminates as an articular nerve to the wrist joint.

The *palmar branch* of the median nerve arises in the lower forearm, pierces the deep fascia immediately above the wrist, crosses the flexor retinaculum superficially and divides into two branches. The lateral branch supplies the skin of the ball of the thumb; the medial branch innervates the palm of the hand.

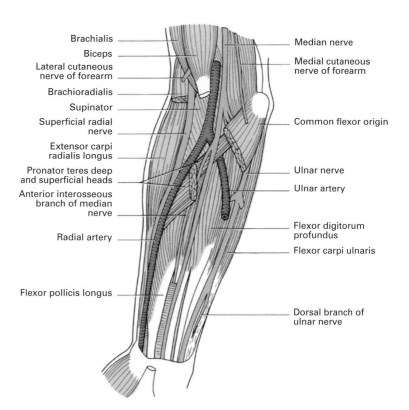

Brachialis

Biceps

Lateral cutaneous nerve of forearm

Brachioradialis

Supinator

Superficial radial nerve

Extensor carpi radialis longus

Pronator teres deep and superficial heads

Anterior interosseous branch of median nerve

Radial artery

Flexor pollicis longus

Median nerve

Medial cutaneous nerve of forearm

Common flexor origin

Ulnar nerve

Ulnar artery

Flexor digitorum profundus

Flexor carpi ulnaris

Dorsal branch of ulnar nerve

Fig. 146 Deep dissection of the forearm.

We might, in passing, list here the five structures that cross the flexor retinaculum superficially:
1 palmaris longus;
2 ulnar artery and its venae comitantes;
3 ulnar nerve;
4 palmar cutaneous branch of the ulnar nerve;
5 palmar cutaneous branch of the median nerve.

The *lateral terminal branch* of the median nerve sends off a short, stout and important branch immediately distal to the flexor retinaculum and 3 cm distal to the distal wrist skin crease, which plunges into the thenar eminence, there to supply the abductor brevis, opponens and flexor brevis of the thumb. This branch is covered only by the thin deep fascia of the thenar eminence and is readily injured by either an accidental cut or an ill-planned surgical incision. The lateral terminal branch then divides into three palmar digital nerves that supply the sides of the thumb and the radial side of the index finger. The last, in addition, innervates the 1st lumbrical muscle.

The *medial terminal branch* divides into two palmar digital nerves that bifurcate to supply adjacent surfaces of the index and middle, and middle and ring fingers, respectively. In addition, the more lateral of the two palmar digital nerves supplies the 2nd lumbrical.

Each digital nerve gives off a collateral branch to the dorsal surface of its terminal phalanx; in this respect, and in the relationship of the nerve to the digital artery, the description given above for the ulnar nerve applies equally well (page 180).

Branches of the posterior cord

The posterior cord of the plexus gives off the upper and lower subscapular nerves and the nerve to latissimus dorsi, then ends by dividing into the axillary and radial nerves.

The *upper* and *lower subscapular nerves* (C5, C6) pass backwards into subscapularis. The lower nerve also supplies teres major.

The *nerve to latissimus dorsi* (*thoracodorsal nerve*) (C6–8) arises between the upper and lower subscapular nerves. It accompanies the subscapular vessels along the posterior axillary wall and supplies latissimus dorsi (Fig. 137).

The *axillary* (*circumflex*) *nerve* (C5, C6), shown in Fig. 147, arises in common with the radial nerve when the posterior cord bifurcates just beyond pectoralis minor. The axillary nerve passes laterally behind the axillary artery and in front of subscapularis. At the lower border of this muscle the nerve passes backwards through the quadrangular space bounded by: subscapularis and teres minor above, teres major below, long head of triceps medially and the surgical neck of the humerus laterally. The nerve is accompanied by the posterior circumflex humeral branches of the axillary vessels; it gives a branch to the shoulder joint, then splits into an anterior and a (larger) posterior division. The anterior division runs around the surgical neck of the humerus a hand's breadth below the acromion; it lies

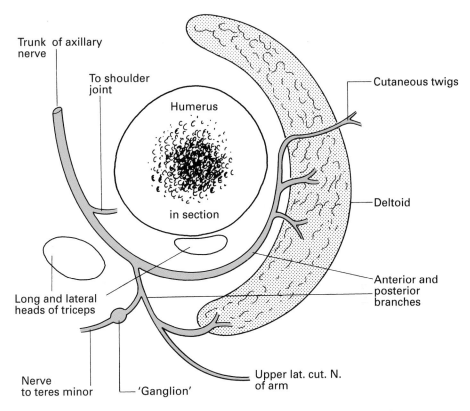

Fig. 147 Schematic section through the upper humerus to show the axillary nerve and its branches. Note the pseudoganglion on the branch to teres minor.

deep to the deltoid, which it supplies. A few filaments pierce the deltoid to reach the overlying skin.

The posterior division supplies teres minor and the posterior part of the deltoid, then curves around the posterior border of this muscle to become the *upper lateral cutaneous nerve of the arm*, supplying the skin over the lower two-thirds of the posterior aspect of the deltoid.

Because of considerable overlap of areas of cutaneous innervation, division of the axillary nerve results in only a small patch of anaesthesia over the lateral aspect of the deltoid – the 'sergeant's patch'. In addition, there is paralysis of the deltoid and therefore a weakness in arm abduction. The *radial nerve* (C5–8, T1), shown in Figs 142 and 148, transmits fibres from all the roots of the brachial plexus. At its origin it lies behind the third part of the axillary artery and descends in turn across subscapularis, teres major and latissimus dorsi; it is easily identified at a dissection of the axilla as it lies on the shining tendinous insertion of this last muscle. The nerve then passes between the long and medial heads of triceps into the posterior compartment of the arm, accompanied by the profunda branches of the brachial vessels. It descends obliquely across the posterior aspect of the humerus along the spiral groove, lying between a superficial muscle plane

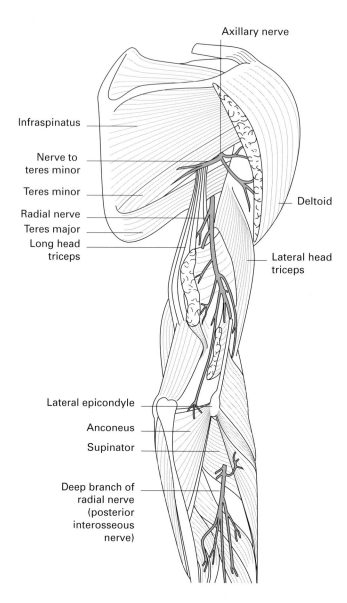

Axillary nerve

Infraspinatus

Nerve to teres minor

Teres minor

Radial nerve

Teres major

Long head triceps

Deltoid

Lateral head triceps

Lateral epicondyle

Anconeus

Supinator

Deep branch of radial nerve (posterior interosseous nerve)

Fig. 148 The distribution of the radial nerve.

formed first by the long and then by the lateral head of the triceps, and a deep plane composed of the medial head of this muscle.

A hand's breadth above the elbow, the nerve reaches the lateral margin of the humerus, pierces the lateral intermuscular septum and enters once more into the anterior compartment of the arm, where it lies between brachialis and brachioradialis. At this point, the nerve is susceptible to compression injury, in particular from an arterial tourniquet placed too low around the arm. It ends in front of the lateral epicondyle of the humerus by dividing into two terminal branches, the superficial radial nerve and the posterior interosseous nerve.

The branches of the radial nerve are muscular, cutaneous, articular (to the elbow) and terminal.

The *muscular branches* are conveniently divided into a medial group (arising in the axilla), a posterior group (arising in the spiral groove) and a lateral group:

1 The medial group supplies:
   a long head of triceps;
   b medial head of triceps.
2 The posterior group supplies:
   a medial head of triceps;
   b lateral head of triceps;
   c anconeus.
3 The lateral group supplies:
   a brachialis (together with the musculocutaneous nerve);
   b brachioradialis;
   c extensor carpi radialis longus.

The nerve to anconeus is in contact with the lower end of the posterior humeral shaft immediately before supplying this muscle; a total of four nerves lie on the humerus: the axillary, radial, ulnar, and the nerve to anconeus. The last, of course, is of no practical importance but all the other three may be, and indeed frequently are, implicated in upper limb injuries (Fig. 142).

The *cutaneous branches* are three in number:

1 The *posterior cutaneous nerve of the arm*, which arises in the axilla and supplies the skin over the proximal one-third of the posterior aspect of the arm.
2 The *posterior cutaneous nerve of the forearm*, which arises in the spiral groove, perforates the lateral head of triceps, descends over the lateral aspect of the forearm to the wrist and supplies the skin over the posterolateral aspect of the forearm.
3 The *lower lateral cutaneous nerve of the arm*, a branch of the latter, also pierces the lateral head of the triceps and supplies an area of skin over the lateral aspect of the arm just above the elbow.

The two *terminal branches* are the posterior interosseous nerve and the superficial radial nerve.

The *posterior interosseous nerve*, apart from articular twigs, is entirely muscular. Arising at the bifurcation of the radial nerve, it plunges into supinator, in which it passes around the lateral side of the radial shaft into the posterior compartment of the forearm. On emerging from the supinator, the nerve breaks up into branches, all of which are muscular apart from one articular strand that descends over the posterior interosseous membrane to the wrist joint.

Its muscular branches are those that arise before the nerve enters supinator, supplying supinator and extensor carpi radialis brevis, and those which arise when the nerve emerges from supinator, supplying extensor digitorum, extensor digiti minimi, extensor carpi ulnaris, extensor pollicis longus, extensor pollicis brevis, extensor indicis and abductor pollicis longus.

The *superficial radial nerve* is entirely sensory. It arises from the bifurcation of the radial nerve deep to brachioradialis. It descends beneath this muscle, lying to the radial side of the radial artery, until about 7.5 cm above the wrist, when it passes backwards under brachioradialis and divides into dorsal digital nerves. These run downwards to supply the dorsal aspect of the thumb base, the radial side of the back of the hand and the backs of the thumb, index, middle and radial half of the ring finger as far as their distal interphalangeal joints.

Variations

The description given of the distribution of the brachial plexus in the hand applies to the most frequent arrangement (Fig. 149), but variations in this pattern are common and important.

On the motor side, the ulnar nerve may encroach on median territory and vice versa. The ulnar nerve may supply part or even all of the thenar muscles, or the median nerve may innervate adductor pollicis and the 1st dorsal interosseous muscle in addition to the three muscles of the thenar eminence. The radial nerve never innervates the intrinsic muscles of the hand.

On the sensory side, either the radial or the ulnar supply to the dorsum of the hand may enlarge at the expense of the other component. The ulnar supply to the dorsum may even, on occasion, be completely deficient. On the palmar aspect, the ulnar nerve may extend its area to innervate the whole of the ring finger or even reach the ulnar side of the middle finger. In other cases, it may be confined to the sensory supply of only the 5th finger.

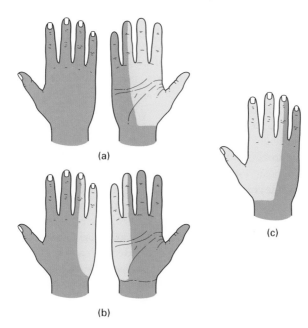

(a)

(c)

(b)

Fig. 149 The normal cutaneous distribution of the nerves of the hand: (a) median; (b) ulnar; (c) radial.

CLINICAL NOTE

Nerve blocks at the elbow and wrist

The ulnar, radial and median nerves can be blocked at the elbow or wrist. These blocks are often performed to supplement patchy brachial plexus blocks, but do not last as long as plexus blocks. They are also occasionally used to provide anaesthesia for operations on the hand, and on occasion for analgesia for dressing changes.

The three nerves passing the elbow can be identified with a nerve stimulator. The ulnar nerve is blocked some 1–3 cm cephalad to the point at which it passes behind the medial epicondyle of the elbow. The median nerve lies just medial to the brachial artery, and the radial nerve lies about 2 cm lateral to the biceps tendon. The cutaneous branches of the musculocutaneous nerve can be blocked by the subcutaneous infiltration of local anaesthetic in the area between the biceps tendon and the lateral epicondyle.

The ulnar, median and radial nerves can also be blocked at the wrist (Fig. 150). Most anaesthetists use a 27G hypodermic needle with a 60° bevel and paraesthesiae to locate the nerves at this level, for fear that large short-bevel stimulating needles may cause trauma to the nerves. The ulnar nerve can be blocked with a needle that passes just lateral to the tendon of flexor carpi ulnaris or, alternatively, by a needle that passes just posterior to it. The median nerve can be blocked with a needle that passes in the midline between the tendons of flexor carpi radialis and palmaris longus, if present. The cutaneous branches of the radial nerve are blocked by a subcutaneous infiltration of local anaesthetic at the base of the thumb.

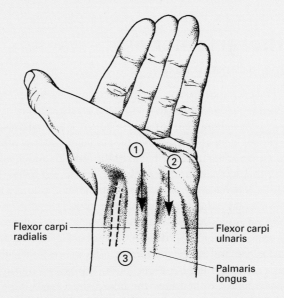

Flexor carpi radialis

Flexor carpi ulnaris

Palmaris longus

Fig. 150 Surface markings of the median nerve (1), ulnar nerve (2) and radial artery (3) at the wrist.

The segmental innervation of the upper limb

The segmental cutaneous supply to the upper limb is shown in Fig. 125. It may be summarized as follows:

1 C3 and C4 supply the upper shoulder region (supraclavicular nerves).
2 C5 supplies the deltoid region and the lateral aspect of the arm.
3 C6 supplies the lateral forearm and the thumb.
4 C7 supplies the hand and the middle three fingers.
5 C8 supplies the 5th finger and medial side on the hand and lower forearm.
6 T1 supplies the medial side of the lower arm and upper forearm.
7 T2 supplies the upper arm on its medial side (via the intercostobrachial nerve).

There is considerable overlap in these cutaneous segments, so that individual block of any one nerve root results in no significant postoperative anaesthesia.

The segmental innervation of the muscles of the upper limb can be grouped thus:

1 C5 supplies the abductors and lateral rotators of the shoulder.
2 C6, C7 and C8 supply the adductors and medial rotators of the shoulder.
3 C5 and C6 supply the elbow flexors.
4 C7 and C8 supply the elbow extensors.
5 C6 supplies the pronators and supinators of the forearm.
6 C6 and C7 supply the long flexors and extensors of the wrist.
7 C7 and C8 supply the long flexors and extensors of the fingers.
8 T1 supplies the intrinsic muscles of the hand.

The thoracic nerves
Anterior primary rami

There are 12 pairs of thoracic anterior primary rami: the upper 11 comprise the intercostal nerves and the 12th is termed the subcostal nerve. They are responsible for the innervation of the muscles of the intercostal spaces and of the anterior abdominal wall, and for the cutaneous supply of the skin of the medial aspect of the upper arm and of the anterior and lateral aspects of the trunk from the level of the angle of Louis to just above the groin. In addition, each nerve is joined to its corresponding thoracic sympathetic ganglion by a white and grey ramus communicans.

The 3rd to 6th intercostal nerves are wholly typical; the others all exhibit variations, to a greater or lesser degree, on this basic plan (Fig. 122).

Intercostal nerves 3–6 (the 'typical' intercostal nerves) enter their intercostal spaces across the anterior aspect of the corresponding superior costotransverse ligament to lie below the intercostal vessels (Fig. 151), first between the posterior intercostal membrane and the pleura and then, at the rib angles, between the internal intercostal and the innermost

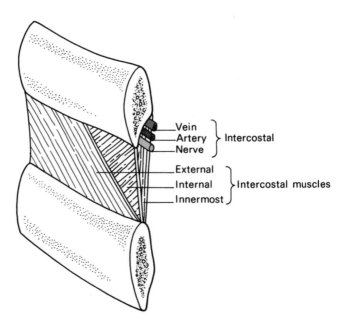

Fig. 151 Diagram of an intercostal space to show the relations of the intercostal nerve.

intercostal muscles. Near the margin of the sternum, each nerve passes in front of the internal thoracic vessels and transversus thoracis (sternocostalis) and pierces the internal intercostal muscle, anterior intercostal membrane (the fibrous anterior part of the external intercostal muscle) and the overlying pectoralis major to become an *anterior cutaneous nerve* of the thorax. Branches from these four 'typical' intercostal nerves are:

1 *muscular*, to the intercostal muscles;
2 *collateral*, which run along the lower border of each intercostal space, and either rejoin the main nerve or end as separate anterior cutaneous nerves;
3 *lateral cutaneous*, which reach the skin in the mid-axillary line and divide into anterior and posterior branches.

The *1st intercostal nerve* (T1) sends a large contribution that passes across the front of the neck of the 1st rib, lateral to the superior intercostal artery, to enter into the composition of the brachial plexus (see page 165). The remaining, and smaller, part of the nerve constitutes the 1st intercostal nerve proper. It has no lateral cutaneous branch and its anterior cutaneous branch, if indeed present at all, is small.

The *2nd intercostal nerve* differs from the 'typical' intercostal nerves only in that its lateral cutaneous branch crosses the axilla to supply the skin over the medial aspect of the upper arm; this branch is termed the *intercostobrachial nerve*. It is inevitably destroyed in the performance of a radical mastectomy, with resultant numbness in this area. The nerve is in danger also in axillary node biopsy.

The *7th to 11th intercostal nerves* enter the abdominal wall between the interdigitations of the diaphragm with transversus abdominis. The 7th and

8th nerves then pass directly into the posterior rectus sheath, pierce rectus abdominis and the anterior rectus sheath, and terminate in the overlying skin. The 9th to 11th nerves travel between transverse abdominis and the internal oblique to reach the posterior rectus sheath, which they penetrate, then traverse rectus and the anterior sheath to reach the surface. The 7th and 8th nerves slope upwards and medially in their short abdominal course, the 9th runs more or less transversely and the 10th and 11th slope downwards; the 10th nerve supplies the region of the umbilicus (Fig. 125).

The branches of the 7th to 11th intercostal nerves are closely comparable to the upper 'typical' intercostals. Motor branches supply the abdominal as well as the intercostal muscles. In addition, each nerve has a collateral (additional anterior cutaneous) and a lateral cutaneous branch. Sensory filaments of the 7th to 11th intercostal nerves supply the periphery of the diaphragm.

The *12th thoracic* (*subcostal*) *nerve* runs along the lower border of the 12th rib below the subcostal vessels, and passes behind the lateral arcuate ligament to run in front of quadratus lumborum behind the kidney and colon. The nerve then passes between transversus abdominis and internal oblique and then has a course and distribution that are similar to the lower intercostal nerves. However, there is one point of difference: the lateral cutaneous branch of the 12th nerve descends without branching to supply the skin over the lateral aspect of the buttock (see Fig. 235). (Note that the subcostal nerve is the motor supply to the pyramidalis, which lies within the lowest part of the rectus sheath.)

In performing a sympathetic denervation of the upper limb (upper dorsal sympathectomy), the surgeon divides the sympathetic chain immediately below the T3 ganglion, then dissects up the chain, dividing all its connections, but carefully preserving the stellate ganglion with its white ramus from T1. In this way, the sympathetic outflow to the upper limb (T2–7) is cut off by preganglionic section, but the main supply to the head and neck via T1 is preserved; by this manoeuvre the rather unsightly Horner's syndrome is avoided. The thoracoscopic approach gives an excellent view of the upper thoracic sympathetic chain and both sides can be operated upon at the same session.

Some sympathetic fibres to the upper limb may leave the chain below the ganglion of T1 and run directly to the brachial plexus. One fairly constant strand from the ganglion of T2 to the 1st thoracic nerve is termed the *nerve of Kuntz*; it must be sought and divided if the sympathectomy is to be complete.

The lumbar plexus

The lumbar plexus is derived from the anterior primary rami of the 1st, 2nd, 3rd and part of the 4th lumbar nerve roots. About 50% of subjects receive an additional contribution from T12. In much the same way as the brachial plexus (see page 165), the lumbar plexus may be pre-fixed, with its lowest contribution from L3, or post-fixed, when it extends to L5.

Formation of the plexus (Fig. 152)

The plexus assembles in front of the transverse processes of the lumbar vertebrae within the substance of the psoas major. L1, joined in 50% of cases by a branch from T12, divides into an upper and lower division. The upper division gives rise to the iliohypogastric and ilio-inguinal nerves; the lower joins a branch from L2 to form the genitofemoral nerve. The rest of L2, together with L3 and the contribution to the plexus from L4, divide into dorsal and ventral divisions. Dorsal divisions of L2 and L3 form the lateral cutaneous nerve of the thigh and L2–4 form the femoral nerve. The ventral branches join into the obturator nerve (L2–4) and, when present, the accessory obturator nerve (L3 and L4).

Summary of branches of the lumbar plexus

- Iliohypogastric: L1
- Ilio-inguinal: L1
- Genitofemoral: L1, L2
- Dorsal divisions:
  - Lateral cutaneous nerve of thigh: L2, L3
  - Femoral nerve: L2–4

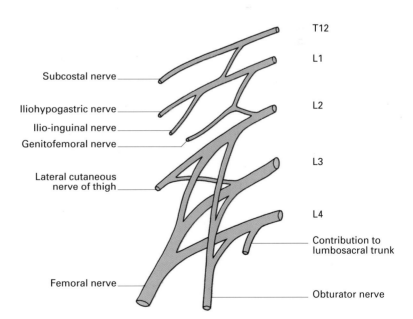

Fig. 152 Plan of the lumbar plexus (muscular branches have been omitted for clarity).

- Ventral divisions
  - Obturator nerve: L2–4
  - Accessory obturator nerve: L3, L4

  In addition, muscle branches are given to:
1 psoas major;
2 psoas minor;
3 iliacus;
4 quadratus lumborum.

The intimate relations of the plexus to the psoas major should be noted (Fig. 153): the obturator nerve, and the accessory obturator when this is present, emerge on its medial border, the genitofemoral pierces the

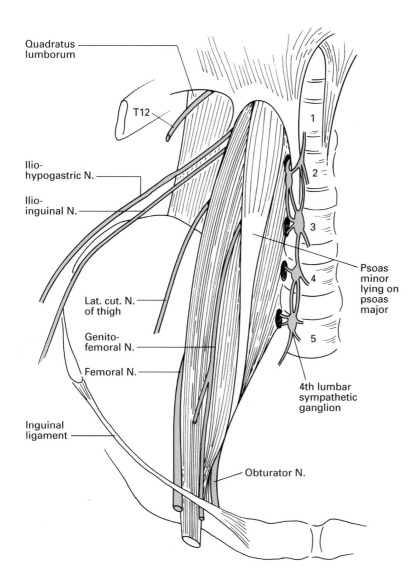

Quadratus lumborum

T12

Ilio-hypogastric N.

Ilio-inguinal N.

Lat. cut. N. of thigh

Genito-femoral N.

Femoral N.

Inguinal ligament

Psoas minor lying on psoas major

4th lumbar sympathetic ganglion

Obturator N.

Fig. 153 The relationships of the lumbar plexus to the psoas.

muscle to lie on its anterior surface, and the remaining nerves appear seriatim along the lateral border.

CLINICAL NOTE

Lumbar plexus block

This block is uncommonly used as an elective anaesthetic tool but can provide complete anaesthesia for hip surgery and for intermittent or continuous analgesia in selected patients. The patient is placed in the lateral position with the knees flexed, the side to be blocked being uppermost. The midline of the back is marked along the spinous processes, as is the line between the most cephalad palpable edges of the iliac crests (the intercristal line). A point approximately 4 cm lateral to the intersection of these two lines is used as the needle entry point. A needle of at least 10 cm length is passed perpendicular to the skin. If it makes contact with a transverse process, it is angled slightly cephalad or caudad so as to pass the process. A nerve stimulator is used; contractions in the quadriceps femoris muscle form the endpoint, creating the spectacle known as the 'dancing patella'. Local anaesthetic can then be injected. Block of the femoral, obturator and lateral cutaneous nerve of the thigh should result. If hip flexion occurs as a result of psoas stimulation, the needle is too deep. Contraction of quadratus lumborum (posterior abdominal wall) indicates that the needle is too lateral. Threading of a catheter into this area is often difficult.

Distribution of the lumbar plexus (Figs 154, 155)

The courses of the iliohypogastric and ilio-inguinal nerves are dealt with on page 320.

The *genitofemoral nerve* (L1, L2) penetrates the psoas to appear on its anterior surface at the level of the lower border of the 3rd lumbar vertebra. It descends over the anterior surface of the psoas, behind the peritoneum and the ureter, and divides into two terminal branches just above the inguinal ligament. This division occasionally occurs near the origin of the nerve, so that a double nerve emerges from the psoas.

The *genital branch* crosses the termination of the external iliac artery to enter the internal inguinal ring. It traverses the inguinal canal within the spermatic cord, supplies the cremaster muscle, then emerges from the external ring to supply the skin over the scrotum and adjacent thigh. In the female, the nerve accompanies the round ligament and supplies the skin over the anterior part of the labium majus and the mons veneris.

The *femoral branch* descends on the external iliac artery, passes under the inguinal ligament, pierces the deep fascia just lateral to the origin of the femoral artery and innervates an area of skin the size of one's hand immediately below the crease of the groin. This nerve is the basis of the cremasteric reflex, which is active in children. Tickling the skin of the upper part of the thigh produces retraction of the testis.

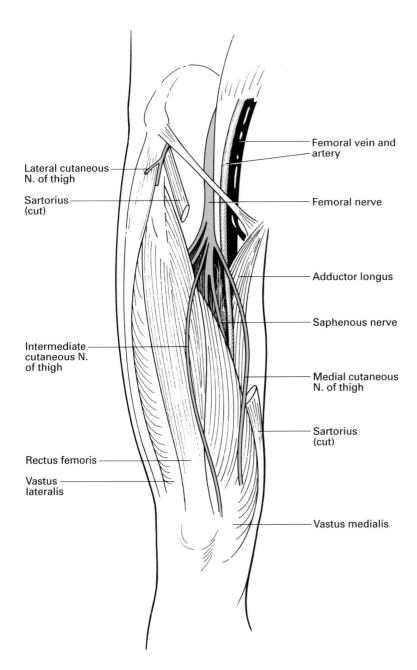

Lateral cutaneous
N. of thigh

Sartorius
(cut)

Femoral vein and
artery

Femoral nerve

Adductor longus

Saphenous nerve

Intermediate
cutaneous N.
of thigh

Medial cutaneous
N. of thigh

Sartorius
(cut)

Rectus femoris

Vastus
lateralis

Vastus medialis

Fig. 154 The
distribution of the
femoral nerve in the
thigh.

The *lateral cutaneous nerve of the thigh* (L2, L3) emerges from the lateral
border of the psoas immediately inferior to the ilio-inguinal nerve. Passing over iliacus, the nerve enters the thigh by running below the lateral
extremity of the inguinal ligament, where it lies on the origin of sartorius and here divides into an anterior and a posterior branch. The anterior

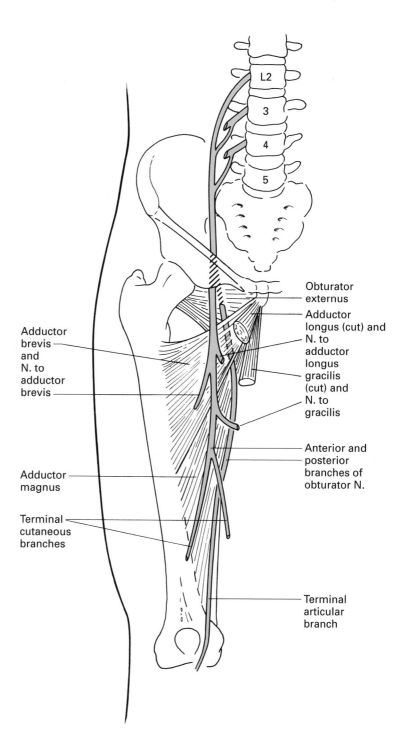

Fig. 155 The distribution of the obturator nerve.

branch supplies the skin over the anterolateral aspect of the thigh down to the knee, where it links up with twigs from the intermediate cutaneous nerve of the thigh and the infrapatellar branch of the saphenous nerve to form the patellar plexus. The posterior branch penetrates the fascia lata to innervate the skin of the lateral aspect of the leg from the greater trochanter to the mid-thigh.

The lateral cutaneous nerve sometimes arises from the femoral nerve and not as a separate branch of the lumbar plexus. It may pierce, rather than pass beneath, the inguinal ligament and may then be compressed between the fasciculi of the ligament with resultant pain, numbness and paraesthesia over the distribution of the nerve (*meralgia paraesthetica*). The pain is relieved by surgical division of the deeper fasciculus of the ligament. Occasionally, similar symptoms are produced by the pressure of an ill-fitting truss or by the involvement of the nerve in its intra-abdominal course by some inflammatory or malignant process. The lateral cutaneous nerve of the thigh can be blocked as it passes medially to the anterior superior iliac spine. The needle entry point is a finger's breadth medial and inferior to the spine. Some 10–15 ml of local anaesthetic deposited in a fan just deep to the inguinal ligament should block the nerve.

The *femoral nerve* (L2–4) is the largest nerve of the lumbar plexus and, in brief, supplies the muscles and the skin of the anterior compartment of the thigh (Fig. 154). The nerve emerges from the lateral margin of psoas, passes downwards in the groove between psoas and iliacus (to both of which it sends a nerve supply), then enters the thigh beneath the inguinal ligament. At the base of the femoral triangle, the nerve lies on iliacus, a finger's breadth lateral to the femoral artery, from which vessel it is separated by a portion of the psoas.

Almost at once within the triangle the nerve breaks up into its terminal branches, which stem from an anterior and posterior division.

Anterior division

- Muscular branches to:
  - pectineus;
  - sartorius.
- Cutaneous branches:
  - intermediate cutaneous nerve of thigh;
  - medial cutaneous nerve of thigh.

Posterior division

- Muscular branches to quadriceps femoris.
- Cutaneous branch – saphenous nerve.
- Articular branches to:
  - hip;
  - knee.

The *nerve to pectineus* passes behind the femoral sheath, in which is contained the femoral artery and vein, and enters the anterior surface of pectineus. This muscle receives in addition an inconstant supply from the accessory obturator nerve.

The *nerve to sartorius* arises either from, or in common with, the intermediate cutaneous nerve of the thigh, and enters the medial aspect of sartorius in its upper third.

The *intermediate cutaneous nerve* of the thigh divides into two branches that supply the front of the thigh down to the knee.

The *medial cutaneous nerve* of the thigh passes medially across the femoral vessels and then divides into anterior and posterior branches. The anterior branch pierces the deep fascia at the lower third of the thigh to supply the skin over the medial side of the lower thigh as far as the knee; here, the nerve links up with the patellar plexus. The posterior branch runs along the posterior border of sartorius, supplying twigs to the overlying skin and communicating with the obturator and saphenous nerves. At the knee, the nerve pierces the deep fascia and supplies an area of skin over the medial side of the leg – an area that is inversely proportional to the contribution from the obturator nerve.

(Note that the lateral, intermediate and medial cutaneous nerves penetrate the deep fascia in echelon, roughly along the oblique line formed by sartorius.)

The *muscular branches of the posterior division* of the femoral nerve supply quadriceps femoris. The nerve to rectus femoris enters the deep aspect of the muscle near its origin; rectus femoris is the only part of the quadriceps to act on the hip as well as the knee and its nerve is the only part of the quadriceps nerve supply to give a branch to the hip joint (see Hilton's law; page 175). The nerve to vastus medialis enters Hunter's canal to supply this muscle. The nerve to vastus intermedius may be bifid or trifid and enters the front of its muscle; a filament descends through intermedius to innervate articularis genu, a distinct part of vastus intermedius inserted into the apex of the synovial membrane of the knee joint, which it pulls upwards during extension of the knee. The nerve to vastus lateralis reaches its muscle by passing deep to rectus femoris in company with the descending branch of the lateral circumflex femoral branch of the profunda femoris artery.

All three nerves to the vasti send filaments of supply to the knee.

The *saphenous nerve* is the largest cutaneous branch of the femoral nerve and the only cutaneous branch to originate from the posterior division. It arises in the femoral triangle, descends lateral to the artery and then enters the adductor canal of Hunter, where it crosses in front of the artery to lie on its medial side. The nerve escapes from the lower part of the canal by emerging between sartorius and gracilis, runs down the medial border of the tibia immediately behind the great saphenous vein, crosses with the vein in front of the medial malleolus and reaches as far as the base of the great toe, supplying an extensive cutaneous area over the medial side of the knee, leg, ankle and foot. The nerve may be inadvertently damaged in exposure of the vein at the ankle.

Immediately on leaving the adductor canal, the saphenous nerve gives off its infrapatellar branch, which pierces sartorius and is distributed to the skin immediately below the knee as part of the patellar plexus.

CLINICAL NOTE

Femoral nerve block and three-in-one blocks

The femoral nerve is relatively easy to block just inferior to the inguinal ligament. A femoral block results in anaesthesia of the entire anterior thigh and most of the femur and knee joint. The block also confers anaesthesia of the skin on the medial aspect of the leg below the knee joint (saphenous nerve) A 50 mm insulated needle is passed 1–2 cm lateral to the femoral artery with a 45° cephalic angulation. Two clicks or pops are felt as the needle passes through the fascia lata and the iliopectineal fascia. Contractions of the quadriceps femoral nerve following stimulation will cause the patella to 'dance or twitch'. Approximately 10–20 ml of local anaesthetic will block the femoral nerve. Larger volumes (30–40 ml) and the use of distal digital pressure may also block the obturator nerve and the lateral cutaneous nerve of the thigh, the three-in-one block. The saphenous nerve can be blocked at the knee. A subcutaneous infiltration of local anaesthetic is performed over the medial side of the knee at the level of the top of the tibia.

The *obturator nerve* (L2–4) emerges from the medial border of psoas at the pelvic brim, which it crosses in its downward and forward passage to the obturator canal (Fig. 155). In this course it passes along the side wall of the pelvis, lateral to the internal iliac vessels and the ureter (Fig. 153). The nerve enters the obturator canal above and anterior to the obturator vessels, which are derived from the internal iliacs.

Within the canal the nerve divides into its anterior and posterior branches.

1 The *anterior branch* passes into the thigh above the obturator externus, descends on adductor brevis, first behind pectineus and then behind adductor longus, to end as a filament that runs along the femoral artery. It supplies adductor longus and gracilis, and frequently adductor brevis; it also sends an articular branch to the hip.

At the lower border of adductor longus, the nerve communicates with the medial cutaneous and saphenous branches of the femoral nerve, forming a so-called subsartorial plexus from which twigs supply the skin of the medial side of the thigh.

2 The *posterior branch* pierces and supplies obturator externus. (Note that obturator internus is supplied from the sacral plexus; see page 204.) It descends on adductor magnus behind adductor brevis, which separates it from the anterior branch of the obturator nerve. The posterior branch supplies adductor magnus and also brevis, if the latter is not served by the anterior branch. Adductor magnus, in addition, receives a supply from the sciatic nerve (see page 209). The nerve then descends along the adductor canal to the popliteal fossa, where it ends by supplying the knee joint.

The *accessory obturator nerve* (L3, L4) is present in about one-third of cases. It appears at the medial border of psoas, crosses the superior pubic ramus, then supplies branches to pectineus and to the hip joint. A communicating branch passes to the anterior division of the obturator nerve.

CLINICAL NOTE

Obturator nerve blocks

Obturator nerve block is sometimes used to treat hip joint pain and in the relief of adductor muscle spasm associated with hemi- or paraplegia. They are becoming less common in anaesthetic practice with the advance of lumbar plexus blocks. The needle entry point is 2 cm below and 2 cm lateral to the pubic tubercle. If the needle is passed directly posteriorly, it will strike the inferior border of the superior ramus of the pubic bone. The needle is then withdrawn slightly and redirected slightly laterally (Fig. 156) so as to walk off the inferior margin of the ramus. It is then advanced a further 2 cm and contractions of the adductor muscles are sought by stimulation to confirm positioning.

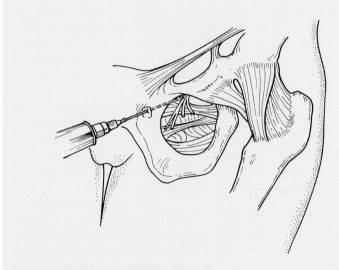

Fig. 156 Obturator nerve block.

The sacral and coccygeal plexuses

There are many variations in the linkages of the sacral and coccygeal anterior primary rami: the following account is that of the usual constitution of these plexuses.

The sacral plexus is formed from a contribution of L4, from the entire L5, S1, S2 and S3 anterior primary rami, and from a part of S4. The coccygeal

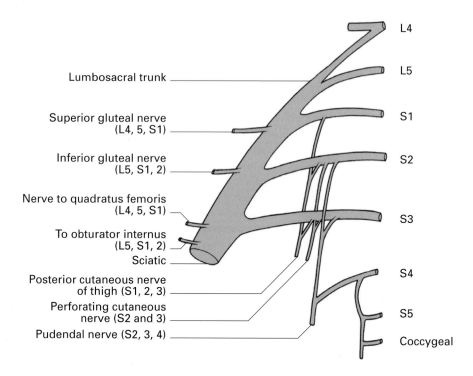

Lumbosacral trunk

Superior gluteal nerve
(L4, 5, S1)

Inferior gluteal nerve
(L5, S1, 2)

Nerve to quadratus femoris
(L4, 5, S1)

To obturator internus
(L5, S1, 2)

Sciatic

Posterior cutaneous nerve
of thigh (S1, 2, 3)

Perforating cutaneous
nerve (S2 and 3)

Pudendal nerve (S2, 3, 4)

L4

L5

S1

S2

S3

S4

S5

Coccygeal

Fig. 157 Plan of the sacral plexus.

plexus receives the rest of S4, together with S5 and the anterior primary ramus of the coccygeal nerve.

Formation of the plexuses (Fig. 157)

The contribution from L4 together with the whole anterior ramus of L5 fuse to form the lumbosacral trunk at the medial border of psoas, medial to the obturator nerve. This trunk passes over the pelvic brim and joins S1 in front of the sacroiliac joint. The anterior primary rami of S1–4 emerge through the anterior sacral foramina, S5 appears between the inferior lateral angle of the sacrum and the transverse process of the coccyx, and Co.1 escapes below this transverse process, piercing the coccygeal muscle.

Relations (Fig. 158)

The roots of the sacral plexus join to form a wide band, which splits to form two terminals, the sciatic and pudendal nerves. The plexus lies on the posterior wall of the pelvic cavity, behind the pelvic fascia and on the anterior surface of piriformis. In front lie the internal iliac vessels and the ureter, together with the sigmoid colon on the left and loops of

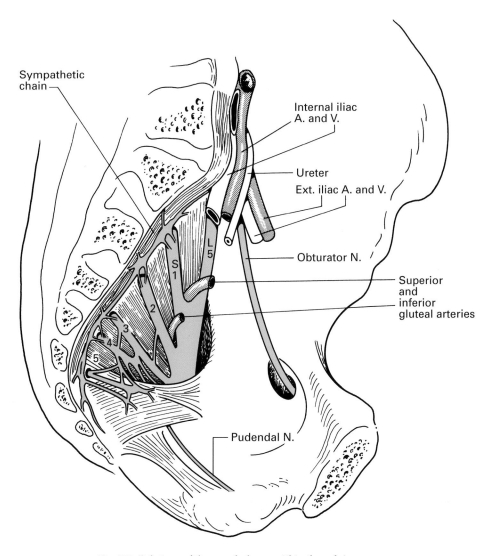

Fig. 158 Relations of the sacral plexus within the pelvis.

lower ileum on the right. The coccygeal plexus lies on the anterior aspect of coccygeus.

Within the pelvic basin the plexus is pierced by four arteries and veins:

1 The iliolumbar vessels pass between the 4th and 5th lumbar roots.

2 The superior gluteal vessels pass between the lumbosacral trunk and S1, or between the roots of S1 and S2.

3 The inferior gluteal vessels pass between the roots of S1 and S2 or of S2 and S3.

4 The internal pudendal vessels pass between the sciatic and pudendal nerves.

Summary of the branches of the sacral plexus

The branches of the sacral plexus are classified into collateral and terminal.
- *Collateral:*
  - muscular;
  - cutaneous;
  - visceral (parasympathetic to the pelvic plexus).
- *Terminal:*
  - sciatic;
  - pudendal.

The collateral branches

The collateral branches and their roots of origin are as follows:
- *Muscular:*
  - nerve to quadratus femoris L4, L5, S1;
  - nerve to obturator internus L5, S1, S2;
  - nerve to piriformis S1, S2;
  - superior gluteal nerve L4, L5, S1;
  - inferior gluteal nerve L5, S1, S2;
  - nerves to levator ani, coccygeus and external anal sphincter S4.
- *Cutaneous:*
  - posterior cutaneous nerve of thigh S1, 2, 3;
  - perforating cutaneous nerve S2, 3.
- *Visceral:*
  - pelvic splanchnics S2, 3.

Muscular collateral branches (Fig. 159)

The *nerve to quadratus femoris* (L4, L5, S1) passes through the lower compartment of the great sciatic foramen, i.e. below piriformis, between the ischium and the deep aspect of the sciatic nerve. It descends over the back of the hip joint, beneath the gemelli and obturator internus tendon, to end by supplying quadratus femoris. In addition, it innervates gemellus inferior and also sends a twig to the hip joint.

The *nerve to obturator internus* (L5, S1, S2) passes into the buttock through the greater sciatic foramen below piriformis, then crosses the ischial spine between the internal pudendal vessels medially and the sciatic nerve laterally. In its brief gluteal appearance it supplies gemellus superior, then passes through the lesser sciatic foramen into the lateral wall of the ischiorectal fossa, there to supply obturator internus.

The *nerve to piriformis* (S1, S2) has a short course to enter the pelvic aspect of piriformis; occasionally, it is bifid.

The *superior gluteal nerve* (L4, L5, S1) accompanies the superior gluteal vessels as the only structures that pass through the upper compartment of the greater sciatic foramen (above piriformis). It supplies gluteus medius and minimus and tensor fasciae lata.

The *inferior gluteal nerve* (L5, S1, S2) passes through the lower compartment of the greater sciatic foramen to enter the deep aspect of gluteus maximus.

Muscular branches from S4 pass from the trunk of S4 to levator ani and coccygeus, which are supplied on their pelvic aspects. In addition, a filament termed the perineal branch of the 4th sacral nerve pierces coccygeus, enters the ischiorectal fossa and descends to supply the external anal sphincter.

Cutaneous collateral branches

The *posterior cutaneous nerve of the thigh* (S1–3) emerges through the greater sciatic foramen below piriformis, on the medial side of the sciatic nerve, and descends over the back of the leg as far as the mid-calf. It gives off:
1 branches to the posterior aspect of the thigh, the popliteal fossa and the upper calf;
2 gluteal branches, which hook around the lower border of gluteus maximus and supply the inferolateral part of the buttock;
3 a perineal branch that passes below the ischial tuberosity forwards to the scrotum (or labium majus).

The *perforating cutaneous nerve* (S2, S3) pierces the sacrotuberous ligament and then hooks around the lower border of gluteus maximus to innervate the skin over the inferomedial aspect of the buttock. The nerve may arise on occasion from either the posterior cutaneous nerve of the thigh or from the pudendal nerve.

Collateral visceral branches

The pelvic splanchnic nerves (S2, S3) are the white rami communicantes that transmit parasympathetic fibres from the roots of the 2nd, 3rd and, occasionally, the 4th sacral nerves to the pelvic autonomic plexuses supplying the pelvic viscera (see page 241).

The terminal branches (Fig. 159)

The *sciatic nerve* (L4, L5, S1–3) is the largest peripheral nerve in the body; at its origin it is a flattened band rather more than 1 cm in width. As befits its size, it transmits fibres from all the roots of the sacral plexus. In reality, it is made up of two nerves, the tibial and common peroneal, within a common sheath of fibrous tissue. Usually, the sciatic nerve splits into these two components at the apex of the popliteal fossa, but the division may occur at any level proximally. Occasionally, the two components are separate right from their origins from the sacral plexus, in which case the common peroneal nerve usually pierces piriformis (10% of cases).

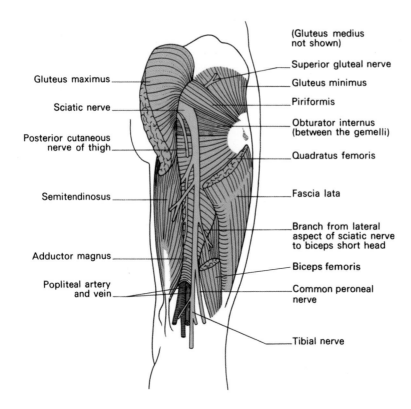

Gluteus maximus

Sciatic nerve

Posterior cutaneous
nerve of thigh

Semitendinosus

Adductor magnus

Popliteal artery
and vein

(Gluteus medius
not shown)

Superior gluteal nerve

Gluteus minimus

Piriformis

Obturator internus
(between the gemelli)

Quadratus femoris

Fascia lata

Branch from lateral
aspect of sciatic nerve
to biceps short head

Biceps femoris

Common peroneal
nerve

Tibial nerve

Fig. 159 Dissection of the sciatic nerve in the thigh and popliteal fossa. Gluteus medius has been removed, revealing the underlying gluteus minimus.

Descending within the substance of the sciatic nerve is the arteria comitans derived from the inferior gluteal artery. In performing an above-knee amputation, this vessel needs to be identified, tied and ligated before the sciatic nerve is divided; tying the vessel may otherwise incorporate nerve fibres and result in severe pain in the stump.

Course

The sciatic nerve leaves the posterior pelvic wall through the greater sciatic foramen below piriformis and enters the region of the buttock very slightly medial to the half-way point between the ischial tuberosity and the greater trochanter. The sciatic nerve then descends vertically down the midline of the back of the leg as far as the apex of the popliteal fossa.

Anteriorly, the nerve rests seriatim on the dorsum of the ischium (with the nerve to quadratus femoris intervening between the sciatic nerve and the bone), gemellus superior, tendon of obturator internus, gemellus inferior, quadratus femoris and, finally, adductor magnus. The upper part of the nerve is under cover of gluteus maximus. Immediately distal to this muscle, the nerve is subfascial and may be injured by a relatively trivial wound. It is then crossed superficially, obliquely and from the medial to the lateral side by the long head of biceps femoris.

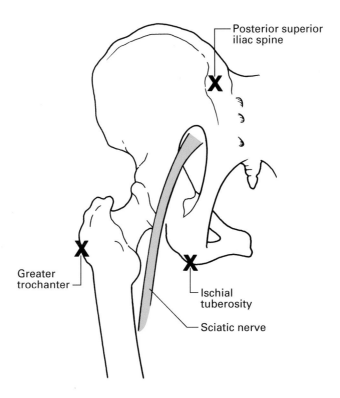

Fig. 160 The surface markings of the sciatic nerve. The mid-point between the ischial tuberosity and the posterosuperior iliac spine is joined to the mid-point between the ischial tuberosity and the greater trochanter by a curved line that is continued vertically down the back of the leg. This line represents the course of the sciatic nerve.

Surface markings (Fig. 160)

The sciatic nerve can be represented by a line that commences at a point mid-way between the posterosuperior iliac spine (easily identified as it lies deep to the sacral dimple) and the tip of the ischial tuberosity, curves outwards and downwards just medial to a point mid-way between the greater trochanter and the ischial tuberosity, and then continues vertically downwards in the midline of the posterior aspect of the thigh.

CLINICAL NOTE

Sciatic nerve blocks

There are a variety of approaches to the sciatic nerve. A popular and reliable one is that described originally by Labat – the posterior approach. With the patient in the lateral position with the knees flexed, a line is drawn between the greater trochanter and the posterior superior iliac spine. At the mid-point of this line, a perpendicular line is drawn. The entry point of the needle is 4–5 cm along this perpendicular. As can be seen from Fig. 160, this will direct the needle towards the sciatic nerve. A nerve stimulator is used; dorsiflexion of the foot will ensue when the needle is in the correct position. Other approaches such as the anterior, lateral and Raj approach are favoured by some, as the patient need not be turned into the lateral position. However, if a lumbar plexus block is to be performed in addition to a sciatic nerve block to provide anaesthesia or analgesia of the whole lower limb, the patient will already be in the ideal position for a Labat sciatic nerve block.

Sciatic nerve damage following injection

It would seem inconceivable that a nerve with such constant and readily defined landmarks could be damaged by intramuscular injections in the buttock, but this accident is in fact so frequently encountered that it has been suggested seriously that the site should be banned. The explanation is, we believe, a psychological one; the standard advice given is to employ the upper outer quadrant of the buttock for these injections, and when the full anatomical extent of the buttock – extending upwards to the iliac crest and outwards to the greater trochanter – is implied, this is perfectly sound advice. Many health-care professionals have an entirely different mental picture of the buttock – a much smaller and more aesthetic affair comprising merely the hillock of the natus. An injection into the upper outer quadrant of this comparatively diminutive structure lies in the immediate neighbourhood of the sciatic nerve. A better surface marking for the 'safe area' of buttock injections can be defined as that area which lies under the outstretched hand when the thumb and thenar eminence are placed along the iliac crest with the tip of the thumb touching the anterosuperior iliac spine. Another useful landmark is to inject anterior to a line joining the greater trochanter to the posterosuperior iliac spine (Fig. 161).

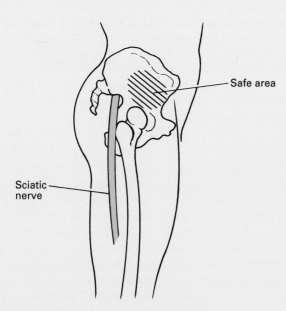

Fig. 161 The 'safe area' for injections into the buttock.

Branches

The branches of the sciatic nerve can be grouped into the following:

1 *Muscular* to:
   a semitendinosus;
   b semimembranosus;
   c adductor magnus;
   d biceps femoris.
2 *Articular* to:
   a the hip joint.
3 *Terminal* to:
   a common peroneal nerve;
   b tibial nerve.

There are a number of features of interest in the arrangement and distribution of the muscular branches. The true hamstring muscles, which have their origin from the ischium, are supplied from the medial side of the sciatic nerve by the tibial component or, indeed, by the tibial nerve itself in those 10% of cases of high division of the nerve. These muscles are the semitendinosus, the semimembranosus and the long head of biceps. In addition, the ischial head of origin of adductor magnus can be considered as a modified hamstring and it too is supplied by the tibial. The true adductor component of adductor magnus, arising from the ramus of the ischium, is supplied by the obturator nerve (see page 200). The short head of biceps, originating from the posterior aspect of the femoral shaft, is developmentally part of the gluteus maximus and not a hamstring; it is the only one of this muscle group to be innervated by the common peroneal component of the sciatic nerve, arising from the lateral side of the nerve.

The *tibial nerve* (L4, L5, S1–3) is the larger of the two terminal branches of the sciatic nerve. It usually arises at the apex of the popliteal fossa (Fig. 162) but, as already noted, it may originate more proximally or may even have a separate origin from the sacral plexus. The nerve traverses the popliteal fossa; it therefore crosses, from above downwards, the popliteal surface of the femur, the posterior aspect of the capsule of the knee joint and the popliteus muscle. Above, the nerve is overlapped by semimembranosus and semitendinosus medially and by biceps femoris laterally; below, it is covered by the two heads of gastrocnemius.

At first the tibial nerve lies to the lateral side of the popliteal vessels – a relationship that is easily explained, since the vessels approach the popliteal fossa somewhat medially through the adductor hiatus. However, the nerve then crosses the vessels superficially to reach their medial side at the lower end of the fossa.

Branches in the popliteal fossa

1 *Muscular* to:
   a popliteus;
   b gastrocnemius;

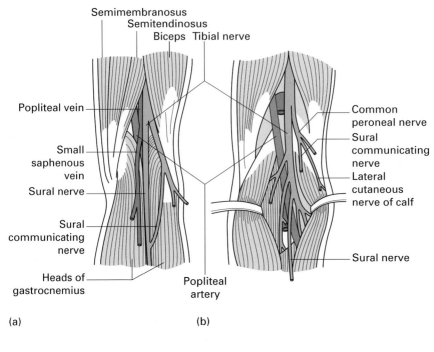

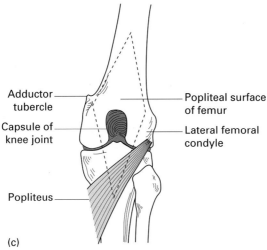

Fig. 162 The right popliteal fossa: (a) superficial dissection; (b) deep dissection; (c) floor.

    c soleus;
    d plantaris.
2 *Cutaneous:*
    a the sural nerve.
3 *Articular:*
    a to the knee.

All the muscular branches of the tibial nerve within the popliteal fossa, apart from that to the medial head of gastrocnemius, are given off from its lateral border (compare with the sciatic nerve; see above).

The sciatic nerve can be blocked in the popliteal fossa by a direct posterior or lateral approach. At this level, the nerve has usually divided into its two terminal branches and, as a result, separate identification and block of the two branches is recommended.

The *sural nerve* (L5, S1, S2) arises in the popliteal fossa (Fig. 162) between the two heads of gastrocnemius, pierces the deep fascia halfway down the posterior aspect of the leg, receives the sural communicating branch of the common peroneal nerve, descends behind the lateral malleolus and then runs along the lateral side of the foot to the 5th toe. It supplies a wide area over the posterolateral part of the lower third of the calf, as well as the lateral side of the foot and of the 5th toe.

The tibial nerve then continues (Fig. 163) distal to the lower margin of the popliteus. Here, the nerve passes under the arch of origin of soleus to descend first on tibialis posterior and then, in the lower leg, directly on the posterior surface of the tibial shaft. Above, the nerve is covered by the muscle bellies of soleus and gastrocnemius; in the lower third of the leg it lies immediately below the deep fascia. It is accompanied through its course by the posterior tibial vessels, which lie first lateral to the nerve, but which cross deep (anterior) to the nerve just below popliteus to descend on its medial side. The complex relationship of the vessels to the tibial nerve and its continuation are readily explained; the nerve itself descends vertically through the popliteal fossa and the leg, but the vessels, at first on the medial side of the tibial nerve, are pulled laterally by the anterior tibial vessels as these pass above the interosseous membrane between the tibia and fibula, then gently swing back again to the medial side of the tibial nerve.

The tibial nerve ends behind the medial malleolus by dividing into its terminal branches – the medial and lateral plantar nerves. Here, the nerve lies beneath the flexor retinaculum and can be marked on the skin by a vertical line drawn halfway between the medial malleolus and the tendo achilles.

It is useful here to recapitulate the order in which the various structures pass behind the medial side of the ankle (Fig. 164); working from the tip of the medial malleolus laterally, these are:

1 tendon of tibialis posterior;
2 tendon of flexor digitorum longus;
3 posterior tibial vein;
4 posterior tibial artery;
5 tibial nerve;
6 tendon of flexor hallucis longus.

Branches of the tibial nerve in the calf and foot

1 *Muscular* to:
  a tibialis posterior;
  b flexor digitorum longus;

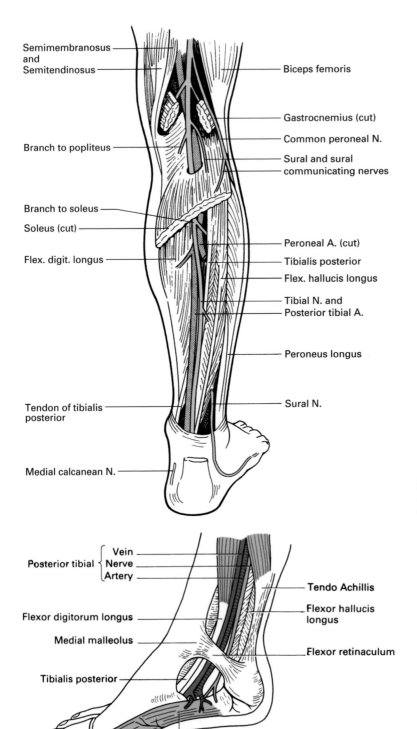

Semimembranosus and Semitendinosus

Biceps femoris

Gastrocnemius (cut)

Common peroneal N.

Branch to popliteus

Sural and sural communicating nerves

Branch to soleus

Soleus (cut)

Peroneal A. (cut)

Tibialis posterior

Flex. digit. longus

Flex. hallucis longus

Tibial N. and Posterior tibial A.

Peroneus longus

Tendon of tibialis posterior

Sural N.

Medial calcanean N.

Fig. 163 The course of the tibial nerve in the calf.

Posterior tibial { Vein / Nerve / Artery

Tendo Achillis

Flexor hallucis longus

Flexor digitorum longus

Medial malleolus

Flexor retinaculum

Tibialis posterior

Abductor hallucis

Fig. 164 The relations of the tibial nerve as it passes behind the medial malleolus.

   c flexor hallucis longus;
   d soleus.
2 *Cutaneous:*
   a the medial calcaneal nerve.
3 *Articular:*
   a to the ankle joint.
4 *Terminal:*
   a medial plantar nerve;
   b lateral plantar nerve.
(Note that the soleus has a double nerve supply from the tibial nerve: one branch enters its superficial surface and a second plunges into its deep aspect.)

The *medial calcaneal nerve* pierces the flexor retinaculum to supply the skin over the medial side of the sole of the foot.

The *medial plantar nerve* is the larger of the two terminal branches of the tibial nerve. Its distribution so closely resembles that of the median nerve in the hand (see page 188 and Fig. 149) that it is only necessary to remember two differences – that the medial plantar supplies only one lumbrical compared with the median nerve's two and that, instead of an opponens muscle, it supplies flexor digitorum brevis.

Arising beneath the flexor retinaculum, the nerve passes deep to abductor hallucis, in company with the medial plantar vessels, to lie between this muscle and flexor digitorum brevis, where it gives off its muscular branches and breaks up into its plantar digital branches.

The medial plantar nerve supplies the following:
1 *Muscular* branches to:
   a abductor hallucis;
   b flexor digitorum brevis;
   c flexor hallucis brevis;
   d 1st lumbrical (from the 1st plantar digital nerve).
2 *Cutaneous* branches:
   a to the medial two-thirds of the sole of the foot and the plantar aspect
      of the medial $3\frac{1}{2}$ toes.

The *lateral plantar nerve* is the smaller terminal branch of the tibial nerve. It closely resembles the distribution of the ulnar nerve in the hand (see page 179).

The nerve lies first under abductor hallucis, then, in company with the lateral plantar vessels, it passes across the sole of the foot to the base of the 5th toe, lying between flexor digitorum brevis (in the first layer of the muscles of the sole) and flexor accessorius (in the second layer). At the lateral side of the foot, the plantar digital branches have their origin; the deep part of the nerve, still accompanied by the vessels, continues back across the sole between adductor hallucis (third layer of muscles) and the interossei (fourth layer).

The lateral plantar nerve supplies the following:
1 *Muscular* branches to:
   a all the interossei;
   b lumbricals 2, 3 and 4;

c adductor hallucis;
d flexor digiti minimi brevis;
e flexor accessorius;
f abductor digiti minimi.

(That is, all the small muscles of the sole of the foot not innervated by the medial plantar nerve.)

2 *Cutaneous* branches to the lateral one-third of the sole of the foot and the plantar aspect of the lateral 1½ toes.

The *common peroneal nerve* (L4, L5, S1, S2) is one of the two terminal branches of the sciatic nerve, and it is but half the diameter of the tibial nerve. It descends from its origin at the apex of the popliteal fossa (Fig. 162) – or higher (see page 209) – obliquely along the medial border of biceps, between this muscle and the lateral head of gastrocnemius.

It then winds round the neck of the fibula, deep to peroneus longus, there to divide into its terminal branches – the deep peroneal and the superficial peroneal nerves (Fig. 165).

Branches (while part of the sciatic nerve; see page 209)

The branches of the common peroneal nerve can be grouped into the following:

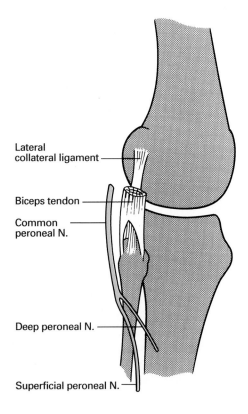

Lateral
collateral ligament

Biceps tendon

Common
peroneal N.

Deep peroneal N.

Superficial peroneal N.

Fig. 165 The common peroneal (lateral popliteal) nerve at the knee. It can be palpated against the neck of the fibula and is the only palpable nerve in the lower limb.

1 *Cutaneous:*
   a sural communicating nerve;
   b lateral cutaneous nerve of calf.
2 *Articular:*
   a to the knee.
3 *Terminal:*
   a deep peroneal (anterior tibial) nerve;
   b superficial peroneal (musculocutaneous) nerve.

The *sural communicating nerve* arises from the common peroneal nerve in the popliteal fossa, and descends over the lateral head of gastrocnemius to join, and be distributed with, the sural nerve (see page 211). Occasionally, it fails to communicate and, as an independent nerve, it then supplies the skin over the lateral side of the leg and ankle.

The *lateral cutaneous nerve of the calf* also arises in the popliteal fossa and also descends over the lateral head of gastrocnemius. It supplies the skin over the anterolateral and posterolateral aspects of the upper calf.

The *deep peroneal nerve* (Fig. 166) arises at the bifurcation of the common peroneal nerve between the neck of the fibula and the peroneus longus. It passes deep to the upper part of extensor digitorum longus to reach the anterior aspect of the interosseous membrane. It descends on this membrane, then on to the lower third of the front of the tibia and finally crosses the front of the ankle joint before breaking up into its terminal branches. At first the nerve lies between extensor digitorum longus and tibialis anterior.

However, extensor hallucis longus arises from the second and third quarters of the shaft of the fibula medial to the extensor digitorum longus and therefore becomes the lateral relationship of the nerve; tibialis anterior remains throughout as the medial relation. At the ankle the nerve is crossed obliquely from laterally to medially by the tendon of extensor hallucis longus – the tendon must do so to reach the great toe.

In its course through the leg, the nerve is accompanied by the anterior tibial vessels. Since these pass above the interosseous membrane at its origin and the nerve hooks round the fibular neck, the nerve will obviously first lie on the lateral side of the vessels upon the interosseous membrane. About the middle of the leg, the vessels swing behind the nerve, but move back again to the medial side of the nerve in the lower third of its course.

The deep peroneal nerve supplies the following:
1 *Muscular* branches to:
   a tibialis anterior;
   b extensor hallucis longus;
   c extensor digitorum longus;
   d peroneus tertius.
2 *Articular* branch:
   a to the ankle joint.
3 *Terminal* branches:
   a medial – cutaneous to the adjacent sides of the 1st and 2nd toes;
   b lateral – motor to extensor digitorum brevis and articular to foot joints.

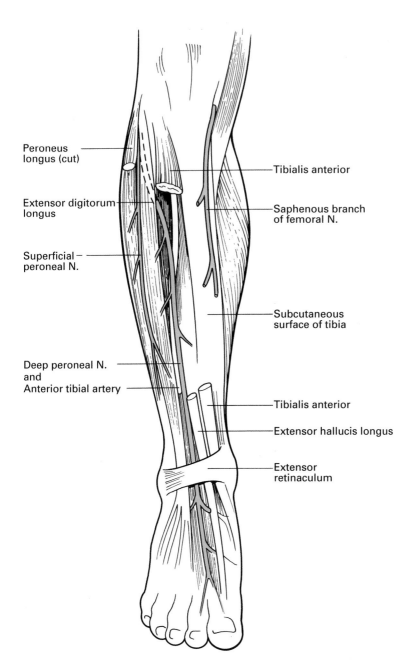

Peroneus longus (cut)

Tibialis anterior

Extensor digitorum longus

Saphenous branch of femoral N.

Superficial peroneal N.

Subcutaneous surface of tibia

Deep peroneal N. and Anterior tibial artery

Tibialis anterior

Extensor hallucis longus

Extensor retinaculum

Fig. 166 Dissection of the deep peroneal nerve and the superficial peroneal nerve at the front of the leg.

The *medial terminal branch* accompanies the dorsalis pedis artery, on the vessel's lateral side, until the latter plunges downwards between the bases of the 1st and 2nd metatarsals. At the web between these toes, the nerve divides to supply the dorsal aspects of the adjacent sides of the 1st and 2nd digits.

(Note that the only cutaneous area supplied by the whole of the deep peroneal nerve is a small patch on the dorsum of the hallux and 2nd toe.)

The *lateral terminal branch* passes deeply to extensor digitorum brevis on the dorsum of the foot, supplies this muscle and then breaks up into filaments to the joints of the foot.

The *superficial peroneal nerve* (Fig. 166) arises in common with the deep peroneal nerve at the bifurcation of the common peroneal nerve on the neck of the fibula. It descends along the intermuscular septum between the peroneal muscles and the extensor group, first with peroneus longus and then brevis laterally and with extensor digitorum longus throughout on its medial side.

The superficial peroneal nerve supplies the following:

1 *Muscular* branches to:
 a peroneus longus;
 b peroneus brevis.
2 *Cutaneous* branches:
 a to the lower outer aspect of the leg.
3 *Terminal* branches:
 a medial – to dorsum of foot and toes;
 b lateral – to dorsum of foot and toes.

The *medial terminal branch* crosses the front of the ankle and then divides. The more medial division runs to the medial side of the hallux; the more lateral splits to supply the adjacent sides of the backs of the 2nd and 3rd toes.

The *lateral terminal branch* supplies the dorsum of the foot, then gives two dorsal digital branches, one to the adjacent sides of the 3rd and 4th toes, the other to the adjacent sides of the 4th and 5th toes.

A recapitulation of the innervation of the dorsum of the toes is thus:

1 sural – lateral side of 5th toe;
2 deep peroneal – adjacent sides of 1st and 2nd;
3 superficial peroneal – the rest.

However, there may be considerable encroachment laterally on the superficial peroneal territory from the sural nerve.

CLINICAL NOTE

Nerve blocks at the ankle

Five nerves pass the malleoli at the ankle: the posterior tibial nerve, the sural nerve, the deep peroneal nerve, the superficial peroneal nerve and the saphenous nerve (Figs 163, 164, 166). All can be blocked with local anaesthetic, although the choice of nerves to be blocked for an individual patient will depend upon the site of surgery.

The posterior tibial nerve is blocked immediately posterior to the medial malleolus as it runs just behind the posterior tibial artery. An injection of 7–10 ml of local anaesthetic at this point will provide anaesthesia, although a smaller volume can be injected if paraesthesiae are elicited during the insertion of the needle. The sural nerve is blocked with a subcutaneous infiltration of local anaesthetic between the lateral malleolus and the tendo achilles. The deep peroneal nerve is blocked with an injection just lateral to the extensor hallucis tendon. The superficial peroneal nerve is blocked by a subcutaneous injection between the extensor hallucis tendon and the lateral malleolus. The saphenous nerve is blocked by a subcutaneous infiltration between the extensor hallucis tendon and the medial malleolus, taking care not to inject local anaesthetic into the saphenous vein.

The *pudendal nerve* (S2–4) provides the principal innervation of the perineum; its course is complex, passing from the pelvis, briefly through the gluteal region, along the side wall of the ischioanal fossa and through the deep perineal pouch to end by supplying the skin of the external genitalia (Figs 158, 167).

Arising as the lower main division of the sacral plexus (although dwarfed by the giant sciatic nerve), the pudendal nerve leaves the pelvis through the greater sciatic foramen below piriformis. It appears briefly in the buttock region, accompanied laterally by the internal pudendal vessels, merely to cross the dorsum of the ischial spine and straightaway disappear through the lesser sciatic foramen into the perineum. The nerve now traverses the lateral wall of the ischiorectal fossa, accompanied by the internal pudendal vessels, and lies within a distinct fascial compartment on the medial aspect of obturator internus termed the pudendal canal

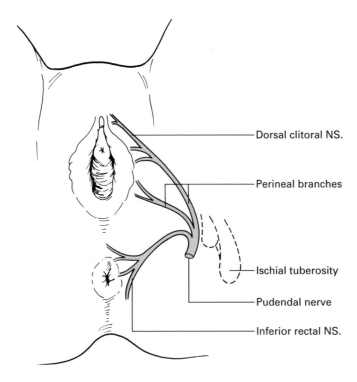

Dorsal clitoral NS.

Perineal branches

Ischial tuberosity

Pudendal nerve

Inferior rectal NS.

Fig. 167 Distribution of the pudendal nerve in the female.

(Alcock's canal). Within the canal, it first gives off the *inferior rectal nerve*, which crosses the fossa to innervate the external anal sphincter and the perianal skin, then divides into the perineal nerve and the dorsal nerve of the penis or clitoris.

The *perineal nerve* is the larger of the two. It bifurcates almost at once; its deeper branch enters the deep pouch and there supplies sphincter urethrae and the other muscles of the anterior perineum – the ischiocavernosus, bulbospongiosus and the superficial and deep transverse perinei. Its more superficial branch innervates the skin of the posterior aspect of the scrotum.

The *dorsal nerve of the penis* (or *clitoris*) traverses the deep perineal pouch, pierces the perineal membrane near its apex, then penetrates the suspensory ligament of the penis to supply the dorsal aspect of this structure.

The sciatic foramina

We might now summarize the boundaries and contents of the greater and lesser sciatic foramina.

The greater foramen is bounded by the margins of the greater sciatic notch and by the sacrotuberous and sacrospinous ligaments; the lesser foramen by the lesser sciatic notch and the same two ligaments (Fig. 168).

The largest structure that emerges through the greater foramen is piriformis, which divides this outlet into an upper and a lower compartment. The upper compartment transmits:
1 the superior gluteal vessels;
2 the superior gluteal nerve.

The lower compartment transmits (from the lateral to medial side):
1 the sciatic nerve; overlying
2 nerve to quadratus femoris; and deep to
3 posterior cutaneous nerve of the thigh;
4 the inferior gluteal nerve;
5 the inferior gluteal vessels;
6 nerve to obturator internus;
7 the internal pudendal vessels;
8 the pudendal nerve.

The three most medial structures (the nerve to obturator internus and the pudendal vessels and nerves) all cross the sacrospinous ligament or ischial spine, then plunge forthwith through the inferior sciatic foramen to enter the perineum. The only other structure transmitted in addition to these by the lesser foramen is the tendon of obturator internus. The five more lateral structures emerging through the greater foramen all cross the dorsum of the ischium and remain in the buttock or descend into the thigh.

The coccygeal plexus

The coccygeal plexus is tiny; made up of a part of S4 together with the whole of S5 and Co.1, it forms a single trunk ('the anococcygeal nerve') that pierces the sacrotuberous ligament to supply the skin over the coccyx.

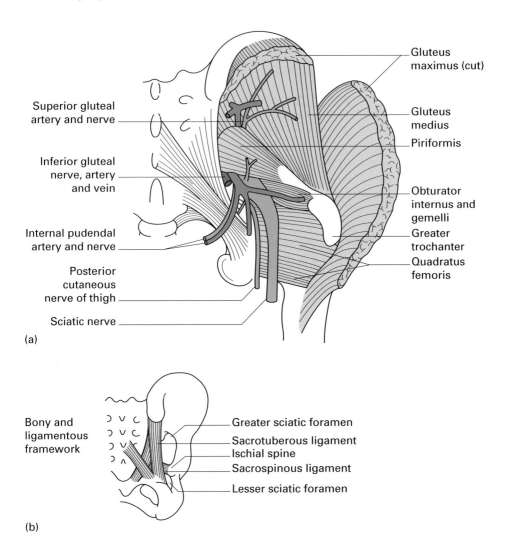

Gluteus
maximus (cut)

Superior gluteal
artery and nerve

Gluteus
medius

Piriformis

Inferior gluteal
nerve, artery
and vein

Obturator
internus and
gemelli

Internal pudendal
artery and nerve

Greater
trochanter

Quadratus
femoris

Posterior
cutaneous
nerve of thigh

Sciatic nerve

(a)

Bony and
ligamentous
framework

Greater sciatic foramen

Sacrotuberous ligament
Ischial spine
Sacrospinous ligament

Lesser sciatic foramen

(b)

Fig. 168 The sciatic foramina. (a) Contents and related muscles. (b) Boundaries and ligamentous framework. Note how piriformis divides the greater sciatic foramen into an upper and a lower compartment.

The segmental innervation of the lower limb

The segmental cutaneous supply to the lower limb is shown in Fig. 125. It may be summarized as follows:

1 L1, L2 and L3 supply the front of the thigh from above down;
2 L4 supplies the anteromedial aspect of the leg;
3 L5 supplies the anterolateral aspect of the leg but also extends on to the medial side of the foot;
4 S1 supplies the lateral side of the foot and the sole;
5 S2 supplies the posterior surface of the leg and thigh;
6 S3 and S4 supply the buttock and perianal region.

(Note that, although S3 supplies the posterior part of the scrotum (or vulva), L1 supplies the anterior part of these structures.)

The segmental innervation of the lower limb muscles is summarized thus:

1 L2 and L3 supply the flexors, adductors and medial rotators of the hip;

2 L3 and L4 supply the extensors, abductors and lateral rotators of the hip;

3 L3 and L4 supply the extensors of the knee;

4 L5 and S1 supply the flexors of the knee;

5 L4 and L5 supply the dorsiflexors of the ankle;

6 S1 and S2 supply the plantar flexors of the ankle;

7 L4 supplies the ankle invertors;

8 L5 and S1 supply the ankle evertors.

Part 5
The Autonomic Nervous System

Introduction

The nervous system can be divided into two great subgroups: the *cerebrospinal system*, made up of the brain, spinal cord and the peripheral cranial and spinal nerves, and the *autonomic system* (also termed the vegetative, visceral or involuntary system), formed by the autonomic ganglia and nerves. Broadly, the cerebrospinal system is concerned with the reactions of the body to the external environment. In contrast, the autonomic system is involved in the control of the internal environment; this is exercised through the innervation of the non-skeletal muscle of the heart, blood vessels, bronchial tree, gut, genitourinary system and pupils, and the secretomotor supply of many glands – those of the alimentary tract and its outgrowths, the sweat glands and, as a special instance, the suprarenal medulla.

These two systems are not to be regarded as independent of each other, for they are, in fact, closely linked both anatomically and functionally. Anatomically, autonomic nerve fibres are transmitted in all of the peripheral and some of the cranial nerves; moreover, the higher connections of the autonomic system are situated in the spinal cord and the brain. Functionally, the two systems are linked within the brain and cord in close physiological integration.

The characteristic feature of the autonomic system is that its efferent nerves emerge as myelinated fibres from the cord or brain, are interrupted in their course by a synapse in a peripheral ganglion and are thence relayed for distribution as fine non-myelinated fibres. In this respect, they differ from the cerebrospinal efferent nerves that pass uninterruptedly to their terminations (Fig. 169).

The autonomic nervous system is subdivided into the *sympathetic* and *parasympathetic* systems on anatomical, functional and, to a considerable extent, pharmacological grounds:

1 *Anatomically*, the sympathetic system has its motor cell station in the lateral grey column of the thoracic and upper two lumbar segments of the spinal cord. The parasympathetic system is anatomically less neatly defined since it is further subdivided into a cranial outflow, along cranial nerves III, VII, IX and X, and a sacral outflow, with cell stations in the 2nd, 3rd and sometimes 4th sacral segments of the cord.

2 *Functionally* (Table 2), the sympathetic system is concerned principally with the stress reactions of the body. Under its influence the pupils dilate; the peripheral blood vessels constrict with consequent shunting of blood to more essential organs; the force, rate and oxygen consumption of the heart increase; the bronchial tree dilates; visceral activity is diminished by inhibition of peristalsis and increase of sphincter tone; glycogenolysis occurs in the liver; the suprarenal medullary secretion is stimulated; and there is pilo-erection and sweating. The sympathetic pelvic nerves inhibit vesical contraction, are motor to the internal vesical sphincter and innervate the uterine musculature.

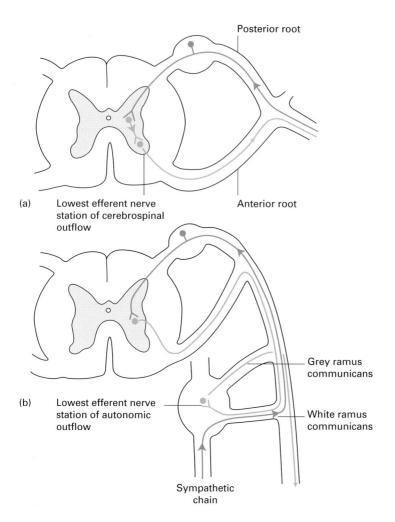

(a) Lowest efferent nerve station of cerebrospinal outflow

Posterior root

Anterior root

(b) Lowest efferent nerve station of autonomic outflow

Grey ramus communicans

White ramus communicans

Sympathetic chain

Fig. 169 The essential difference between the cerebrospinal and autonomic outflows: (a) the cerebrospinal system has its lowest efferent nerve cell stations within the central nervous system; (b) the autonomic system has its lowest efferent cell stations in a peripheral ganglion (here illustrated by a typical sympathetic nerve ganglion).

Although coronary artery blood flow is increased, it is conjectural whether this is a direct sympathetic effect on the coronary arteries or whether it is the result of the following indirect factors: more vigorous myocardial contraction, reduced systole, relatively increased diastole and an increased local concentration of vasodilators or metabolites, all of which would increase coronary flow.

The parasympathetic nerves tend to be antagonistic to the sympathetic system. Thus, stimulation results in pupillary constriction; diminution in the rate, conduction and excitability of the heart; increase in gut peristalsis; relaxation of sphincters; and an increase in alimentary glandular secretion. In addition, pelvic parasympathetic fibres are inhibitors to the vesical internal sphincter and motor to the detrusor muscle of the bladder.

The sympathetic system tends to have a 'mass action' effect; stimulation of any part of it results in a widespread sympathetic response. In contrast, parasympathetic activity is usually discrete and localized.

Table 2 Summary of the effects of sympathetic and parasympathetic stimulation

	Sympathetic stimulation	Parasympathetic stimulation
Eye	Pupil dilates	Pupil constricts; accommodation of lens
Lacrimal gland	Vasoconstrictor	Secretomotor
Heart	Increase in force, rate, conduction and excitability	Decrease in force, rate, conduction and excitability
Lung	Bronchi dilate	Bronchi constrict; secretomotor to mucous glands
Skin	Vasoconstrictor Pilo-erection Secretomotor to sweat glands	–
Salivary glands	Vasoconstrictor	Secretomotor
Musculature of the alimentary canal	Peristalsis inhibited	Peristalsis activated; sphincters relax
Acid secretion of the stomach	–	Secretomotor
Pancreas	–	Secretomotor
Liver	Glycogenolysis	–
Suprarenal	Secretomotor	–
Bladder	Detrusor inhibited Sphincter stimulated	Detrusor stimulated Sphincter inhibited
Uterus	Uterine contraction Vasoconstriction	Vasodilatation

This difference, as will be described below, can be explained, at least in part, by differences in the anatomical peripheral connections of the two systems.

It is perhaps better to think of the sympathetic and parasympathetic systems as synergists rather than antagonists. For example, reflex slowing of the heart is effected partly from increased vagal and partly from decreased sympathetic stimulation. Moreover, some organs receive autonomic innervation from one system only; for example, the suprarenal medulla and the cutaneous arterioles receive only sympathetic fibres, whereas neurogenic gastric secretion is entirely under parasympathetic control via the vagus nerve (X).

3 *Pharmacologically,* the sympathetic postganglionic terminals release epinephrine and norepinephrine; with the single exception of the terminals to the sweat glands which, in common with all the parasympathetic postganglionic terminations, release acetylcholine.

Autonomic afferents

As well as the efferent system considered here, there are *afferent autonomic fibres* that are concerned with the afferent arc of autonomic reflexes and with the conduction of visceral pain stimuli. These nerves have their cell stations in the dorsal root ganglia of the spinal nerves or the ganglia of the cranial nerves concerned with the autonomic system. The fibres from the viscera ascend in the autonomic plexuses; those from the body wall are conveyed in the peripheral spinal nerves. The afferent course from any structure is therefore along the same pathway as the efferent autonomic fibres that supply the part.

The afferent fibres ascend centrally to the hypothalamus and thence to the orbital and frontal gyri of the cerebral cortex along unknown pathways. Normally, we are unaware of the afferent impulses unless they become sufficiently great to exceed the pain threshold when they are perceived as visceral pain, e.g. the pain of coronary ischaemia, intermittent claudication or intestinal colic.

It is now necessary to describe the sympathetic and parasympathetic components of the autonomic system in greater detail.

The sympathetic system

It is convenient to consider this system first at its spinal level, then in its peripheral distribution and finally at its central connections.

Spinal level

The efferent fibres from the central nervous system arise in the lateral grey column of the spinal cord (Fig. 169) from segments T1–L2. From each of these segments, small medullated axons emerge into the corresponding anterior primary ramus and pass via a white ramus communicans into the sympathetic trunk.

Which spinal segments are responsible for the sympathetic innervation of the various regions of the body is not accurately known, but Table 3 is at least an approximation of the truth.

The sympathetic trunk

The sympathetic trunk on each side is a ganglionated nerve chain that extends from the base of the skull to the coccyx, in close relationship to the

Table 3 Segmental distribution of sympathetic fibres

Zone	Spinal segments
Head and neck	T1–2
Upper limb	T2–5
Thoracic viscera	T1–4
Abdominal viscera	T4–L2
Pelvic viscera	T10–L2
Lower limb	T11–L2

vertebral column, maintaining a distance of about 2.5 cm from the midline throughout its course.

Commencing in the superior cervical ganglion beneath the skull base, the chain descends closely behind the posterior wall of the carotid sheath in front of the transverse processes of the cervical vertebrae (Fig. 37). The chain enters the thorax anterior to the neck of the 1st rib, descends over the heads of the upper ribs and then lies on the sides of the bodies of the last three or four thoracic vertebrae. It is covered within the chest by pleura and crosses in front of the intercostal vessels at each intervertebral space (Figs 54, 55).

The chain passes into the abdomen behind the medial arcuate ligament (Fig. 62) and lies in a groove between psoas major and the sides of the lumbar vertebral bodies. It lies in front of the lumbar arteries but may be crossed by the lumbar veins. The left chain is overlapped by the abdominal aorta, the right by the inferior vena cava. The chain then passes behind the common iliac vessels and enters the pelvis anterior to the ala of the sacrum; thence it descends medial to the anterior sacral foramina. The sympathetic trunks end below by meeting each other at the ganglion impar on the anterior face of the coccyx.

The sympathetic trunks bear a series of ganglia that contain motor cells with which preganglionic medullated fibres enter into synapse and from which non-medullated postganglionic axons originate. Developmentally, there was originally one ganglion for each peripheral nerve but by a process of fusion these have been reduced, in humans, to three cervical, 12 or fewer thoracic, two to four lumbar and four sacral ganglia. Only the ganglia T1–L2 receive white rami directly; the higher and lower ganglia must receive their preganglionic supply from medullated nerves that travel through their corresponding ganglion without relay and that then ascend or descend in the sympathetic chain. Still other preganglionic fibres pass through the ganglia intact and pass to peripheral visceral ganglia for relay. There are thus three fates that might befall the white rami (Fig. 170):

1 to enter into synapse in the corresponding sympathetic ganglion (which can only occur in T1–L2 segments);
2 to ascend or descend in the sympathetic chain with relay in higher or lower ganglia;
3 to traverse the ganglia intact and relay in peripheral ganglia; stimulation of a single white ramus communicans will thus obviously have widespread effects – the anatomical basis of the 'mass action' response of sympathetic stimulation.

The branches from the sympathetic ganglionic chain are divided into somatic and visceral groups.

Somatic

Each spinal nerve receives one or more grey rami from a sympathetic ganglion. The grey rami carry postganglionic non-medullated fibres, which are distributed to the segmental skin area supplied by the spinal nerve. These fibres are vasoconstrictor to the skin arterioles, sudomotor to sweat

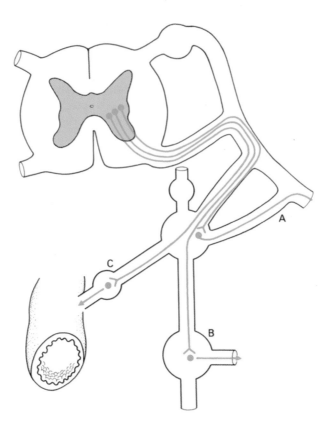

Fig. 170 The three fates of sympathetic white rami. These may: (A) relay in their corresponding ganglion and pass to their corresponding spinal nerve for distribution; (B) ascend or descend in the sympathetic chain and relay in higher or lower ganglia; or (C) pass without synapse to a peripheral ganglion for relay.

glands and pilomotor to the arrectores pilorum, which make the hairs of the skin stand on end.

Visceral

Postganglionic fibres to the head and neck and to the thoracic viscera arise from the ganglion cells of the sympathetic chain. Those to the head ascend along the internal carotid and vertebral arteries; those to the thoracic organs descend to, and are distributed by, the cardiac, pulmonary and oesophageal plexuses.

The abdominal and pelvic viscera, however, are supplied in quite a different manner by postganglionic fibres that have their cell stations in peripheral ganglia – the coeliac, hypogastric and pelvic plexuses – which receive their preganglionic fibres from the splanchnic nerves.

The suprarenal medulla has a unique nerve supply. A rich plexus of preganglionic fibres passes without relay from the coeliac ganglion to the gland; the preganglionic terminals end in direct contact with the chromaffin medullary cells, and liberate acetylcholine (as at all autonomic ganglia), which stimulates the secretion of epinephrine and norepinephrine by the suprarenal medulla. The chromaffin cells of the suprarenal medulla may thus be regarded as sympathetic cells that have not developed postganglionic fibres; indeed, embryologically both the medulla and the sympathetic nerves have a common origin from the neural crest.

The ganglia of the sympathetic trunk

The *cervical ganglia* (Fig. 171) are three in number; they receive preganglionic fibres from spinal segments T1–5 and are the sites of relay for postganglionic fibres to the head, neck and upper limb. These fibres are distributed either along the peripheral spinal nerves or as plexuses around the carotid artery and its branches and the vertebral artery.

The *superior cervical ganglion* is 2.5 cm or more in length and lies opposite the 2nd and 3rd cervical vertebrae; it represents the fused ganglia of C1–4. Its branches are distributed thus:

1 Grey rami communicantes pass to the upper four cervical nerves.

2 Branches surround the internal carotid artery to form the internal carotid plexus, from which they are distributed:

    a as the deep petrosal nerve to the pterygopalatine ganglion (see page 264);

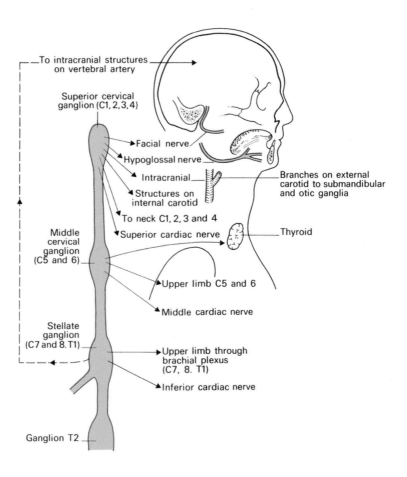

Fig. 171 The cervical sympathetic chain.

b as a root to the ciliary ganglion (see page 261) to supply dilator pupillae;

c as fibres that supply the cerebral vessels and the pituitary.

3 Fibres form a plexus on the external carotid artery and its branches along which they reach the submandibular ganglion (see page 273) and the otic ganglion (see page 274). These fibres are vasomotor to the salivary glands.

4 Grey rami pass to the VIIth, IXth, Xth and XIIth cranial nerves.

5 The superior cardiac nerve descends on the left side to the superficial cardiac plexus and on the right side to the deep plexus.

The *middle cervical ganglion* is small and not always present; it lies at the level of the 6th cervical vertebra and represents the fused ganglia of C5 and C6. Its branches are:

1 grey rami, which join the anterior primary rami of C5 and C6;

2 a thyroid branch, which travels along the inferior thyroid artery to this gland;

3 the middle cardiac nerve, which descends to the deep cardiac plexus.

The *inferior cervical ganglion* lies at the level of the disc space between the 7th cervical and 1st thoracic vertebrae; in 80% of subjects it is fused with the 1st thoracic ganglion to form the *stellate ganglion*. It represents the coalescence of the 7th and 8th cervical ganglia.

The inferior cervical ganglion is connected with the middle ganglion not only by the sympathetic chain itself, but also by the ansa subclavia, which reaches the middle ganglion by looping around the inferior margin of the subclavian artery and then passing upwards in front of it. Its branches are:

1 grey rami to the 7th and 8th cervical nerves;

2 a plexus along the vertebral artery to the brain;

3 the inferior cardiac nerve, which descends to form part of the deep cardiac plexus.

The *thoracic ganglia* (Figs 54, 55) are usually 12 in number, although this number may be decreased by fusion; the commonest example of this is the blending of T1 with the inferior cervical ganglion to form the stellate ganglion. Each ganglion is connected to its corresponding intercostal nerve by grey and white rami communicantes. These branches are the following:

1 Grey rami to intercostal nerves.

2 Branches from T2, T3 and T4 to the cardiac, posterior pulmonary and oesophageal plexuses.

3 Fibres to the wall of the aorta.

4 The splanchnic nerves, which originate from the lower eight ganglia thus:

a the *greater splanchnic nerve* arises from T5–9 (or T10) and passes obliquely downwards and forwards on the sides of the vertebral bodies close to the lateral side of the vena azygos or hemiazygos. It pierces the crus of the diaphragm to join the coeliac ganglion;

b the *lesser splanchnic nerve* (T9 and T10 or T10 and T11) also pierces the crus and ends in the coeliac ganglion;

c the *lowest splanchnic nerve* arises from the lowest available thoracic ganglion and either pierces the crus or passes behind the medial arcuate ligament to join the renal plexus.

The *lumbar ganglia* (Fig. 172) are usually four in number; the upper two receive white rami communicantes from the corresponding lumbar nerves. The branches are:

1 grey rami to the lumbar nerves;

2 branches to the aortic plexus;

3 fibres that descend over the common iliac vessels to the hypogastric plexus.

The ganglia of the lumbar sympathetic chain lie on the anterolateral surface of the bodies of the lumbar vertebrae in loose retroperitoneal areolar tissue separated from the vertebral bodies by the thick anterior ligament. They are overlapped by the aorta on the left and the inferior vena cava on the right (Fig. 173).

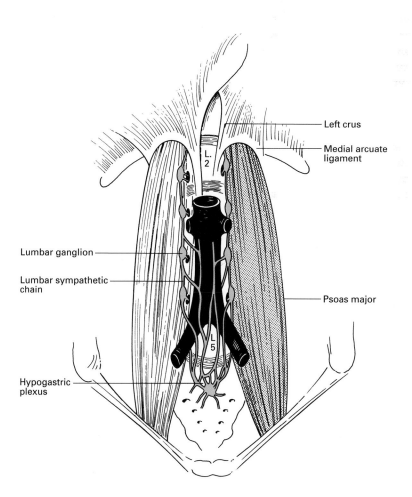

Fig. 172 The lumbar sympathetic chain.

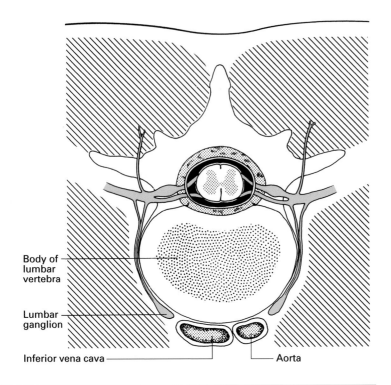

Body of lumbar vertebra

Lumbar ganglion

Inferior vena cava

Aorta

Fig. 173 The relationships of the lumbar sympathetic chain.

CLINICAL NOTE

Lumbar sympathetic block/sympathectomy (Fig. 174)

The lumbar sympathetic ganglia may be blocked temporarily by local anaesthetic in a lumbar sympathetic block (LSB) or semipermanently (or for months) using phenol, alcohol or, more recently, a radiofrequency lesion, when it is termed a lumbar sympathectomy. Sympathetic blockade has been used to treat a variety of disorders:

- symptomatic vasospastic disorders;
- inoperable arterial occlusive disease producing pain at rest, limited ulceration or superficial gangrene;
- complex regional pain syndromes I and II of the lower extremity;
- phantom limb and stump pain;
- pain from acute herpes zoster and post-herpetic neuralgia in lower limb dermatomes;
- acute and chronic renal colic.

LSB is often performed as a predictive procedure prior to sympathectomy, though it has been used in the short-term management of painful conditions. The response to the block is not proportional to changes in perfusion, and alterations in blood flow return to normal over the ensuing months.

Lumbar sympathectomy is usually performed under radiographic image intensification, although computed tomography (CT) guidance has been

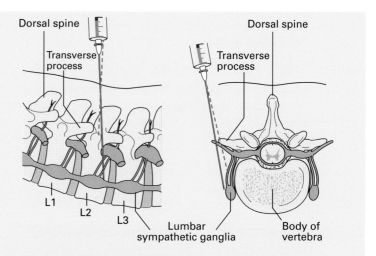

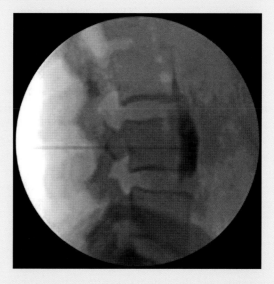

Fig. 174 Lumbar sympathetic block. Diagram showing a needle at the anterolateral aspect of the L2 vertebra adjacent to the sympathetic ganglia.

used. The patient is placed in the lateral or prone position, and a needle is inserted approximately 10–12 cm lateral to the midline. The needle is advanced towards the body of the 2nd or 3rd lumbar vertebra (L2/L3). If the needle touches bone it is withdrawn and is angled more anteriorly, such that it slides off the anterolateral surface of the vertebral body; under image intensification, the needle is advanced to the anterolateral border of the vertebrae. A faint 'pop' may be felt as it crosses the psoas fascia to enter the retroperitoneal space. After aspiration to exclude intravascular placement, 0.5–2.0 ml radio-opaque dye is injected under real-time imaging (Fig. 175). Injection should be easy and imaging should confirm that the

Fig. 175 Lateral lumbar spine radiograph showing contrast injected during a lumbar sympathetic block.

dye spreads up and down the vertebral column just anterior to the body of the lumbar vertebrae. Local anaesthetic (15–20 ml 0.25% bupivacaine with epinephrine) can then be injected. A neurolytic block uses approximately 10 ml of 10% phenol with contrast medium. Opinions vary but there is no evidence that multiple level injections are more effective than a single injection for a neurolytic block. Radiofrequency lesioning of the lumbar sympathetic chain is performed at the L2, L3 and L4 levels.

The *sacral ganglia* are usually four; they send grey rami to the sacral nerves and to the pelvic plexuses.

The plexuses of the sympathetic system

The sympathetic distribution to the intrathoracic and intra-abdominal viscera is via a series of plexuses, termed the cardiac, coeliac and hypogastric, in which nerve relay takes place. Parasympathetic fibres are also transmitted along the branches of these plexuses but relay in ganglion cells situated in the walls of the organs themselves.

The *cardiac plexus* is divided into a superficial and a deep part that have close communications with each other.

The *superficial plexus* lies in front of the pulmonary artery, sheltered within the curve of the aortic arch. It receives the superior cardiac nerve of the left superior cervical sympathetic ganglion and the lower cardiac branch of the left vagus. Branches pass to the deep cardiac plexus, to the left anterior pulmonary plexus and to the plexus along the right coronary artery.

The *deep cardiac plexus* lies in front of the tracheal bifurcation behind the aortic arch. It is made up of the cardiac branches of all the right cervical sympathetic ganglia and of the middle and inferior ganglia on the left, together with branches from the upper four thoracic ganglia and cardiac branches from both vagi (see page 288).

Branches pass to the *pulmonary plexuses* at the lung hila and along the plexuses that accompany both the right and left coronary arteries.

The *coeliac plexus* (Fig. 176) is the largest sympathetic plexus and surrounds the root of the coeliac artery at the level of L1. It consists of a dense felt of fibres uniting into the right and left coeliac ganglia, which are each about 2.5 cm in diameter and which lie on the crura of the diaphragm. The right ganglion is overlapped by the inferior vena cava, the left by the pancreas and the splenic artery. The plexus receives the greater and lesser splanchnic nerves (see page 232) and the coeliac branch of the right vagus.

A large contribution of preganglionic fibres passes from the coeliac plexus to the medulla of the suprarenal gland without relay. The rest of the plexus descends over the abdominal aorta as the *aortic plexus* (which receives contributions from the lumbar ganglia) and is distributed along the branches of the aorta, taking from them the corresponding names of

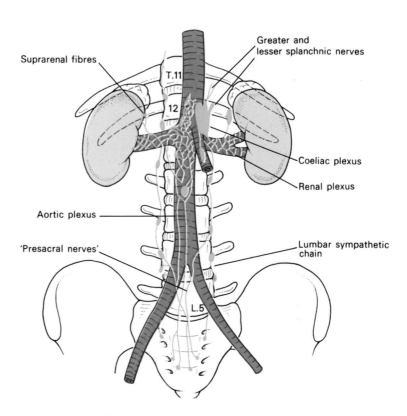

Fig. 176 The coeliac plexus.

the phrenic, hepatic, splenic, left gastric, renal, mesenteric and testicular (or ovarian) plexuses.

The *hypogastric plexus* lies on the promontory of the sacrum and the left common iliac vein, in the cleft between the common iliac arteries. It is formed by the 'presacral nerves' that descend from the aortic plexus and from the lumbar sympathetic trunks. From this plexus, fibres descend along the internal iliac arteries on either side to form the right and left *pelvic plexuses*. They are joined by the pelvic splanchnic nerves, which are the parasympathetic contribution from the 2nd and 3rd (and sometimes 4th) sacral nerves, and are distributed to the pelvic organs.

CLINICAL NOTE

Coeliac plexus block (Fig. 177)

Local anaesthetic block of the coeliac plexus may be used in acute and chronic pancreatitis, or in conjunction with nerve block of the lower intercostal nerves to produce regional anaesthesia for intra-abdominal surgery. Neurolytic coeliac plexus block is often used for the treatment of pain caused by upper gastrointestinal tract malignancy. A coeliac plexus block may interrupt pain-conducting sympathetic afferents. Because of the

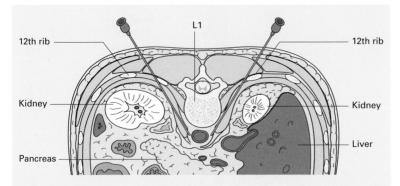

Fig. 177 Diagram showing a coeliac plexus block. Note the close proximity between the needles and the major blood vessels and other important structures.

visceral sympathetic block that is produced and the consequent vasodilatation, some degree of hypotension may be anticipated and intravenous fluids are usually given in advance of the block. The block is performed using radiographic image intensification with the patient lying prone, with intravenous sedation and local anaesthetic infiltration of superficial layers, and normally involves bilateral injections. It may also be performed under CT guidance.

A line is drawn across the patient's back at the level of the 2nd lumbar vertebra. The needles are inserted approximately 8–10 cm from the midline on this line around the 12th rib, ensuring that they pass below the level of the 12th rib. The needles are directed towards the body of the 1st lumbar vertebra at an angle of approximately 45° to the coronal plane and angled slightly cephalad. Once contact has been made with the vertebral body, the needle is withdrawn and then advanced to the anterolateral edge of the vertebral body. If the needle on the left-hand side is in the correct position, pulsation of the aorta is readily transmitted to it. The right-hand needle is advanced slightly further than this. The left-hand needle tip should sit just posterior to the aorta and the right-hand needle tip just anterolateral to the aorta. Aspiration is performed to ensure that the needles are not lying within a vessel.

Injection of radio-opaque dye should demonstrate localization of the dye around the L1 region, with a smooth posterior edge corresponding to the psoas fascia. For local anaesthetic blocks, 10 ml of solution is injected on each side. For neurolytic blocks, phenol or 50% alcohol is used. After a bilateral block, patients should be warned that they might develop temporary postural hypotension. A transaortic technique has also been described.

The diffuse nature of the coeliac plexus makes it less amenable to radiofrequency lesioning than other nerves. Radiofrequency lesioning of the splanchnic nerves under image intensification is increasingly described as an alternative to neurodestructive procedures.

Higher sympathetic centres

Higher centres influencing the sympathetic outflow are located at the brainstem, hypothalamic and cerebrocortical levels. The brainstem centres are situated close to the midline in the floor of the pons and 4th ventricle, and are best represented by the vasomotor centre. The cortical and hypothalamic centres are closely linked and are termed the *limbic system*, which is made up of the gyrus cinguli, hippocampal gyrus and uncus on the medial aspect of the cortex, the anterior thalamic nucleus, amygdala and the hypothalamus. Stimulation of the limbic system is followed by marked changes in heart rate, bowel motility and pupil reactions. Moreover, fibres can be traced from the lateral grey column of the cord through the medulla to the limbic area, which represents the highest level of control of autonomic function.

The parasympathetic system

As already stated, this system has a cranial and a sacral component. Its medullated preganglionic fibres enter into synapse with ganglion cells situated close to, or actually in the walls of, the viscera supplied. Therefore, postganglionic fibres have only a short and direct course to their effector cells, and there is thus the anatomical pathway of a local discrete response to parasympathetic stimulation (Fig. 178).

The cranial outflow

The cranial component of the parasympathetic system is conveyed in cranial nerves III, VII, IX and X, of which X (the vagus) is the most important and most widely distributed. The functions of this group of nerves can be summarized thus:

1 pupils – constrictor to pupil, motor to ciliary muscle (accommodation);
2 salivary glands – secretomotor;
3 lacrimal glands – secretomotor;
4 heart – inhibitor of cardiac conduction, contraction, excitability and impulse formation (with consequent slowing of the heart and a decrease in the force of contraction);
5 lungs – bronchoconstrictor, secretomotor to mucous glands (perhaps vasodilator to blood vessels);
6 alimentary canal – motor to the gut muscles as far as the region of the splenic flexure; inhibitor to the pyloric sphincter. Secretomotor to the glands and adnexae of the stomach and intestine.

The parasympathetic distribution of III, VII and IX is carried out via four ganglia from which the postganglionic fibres relay. These ganglia also transmit (without synapse and therefore without functional connection) sympathetic and sensory fibres, which have similar peripheral distribution.

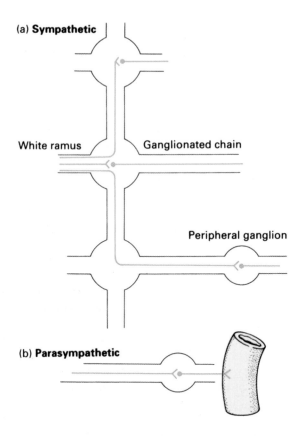

(a) **Sympathetic**

White ramus   Ganglionated chain

Peripheral ganglion

(b) **Parasympathetic**

Fig. 178 The anatomical basis of widespread sympathetic and local parasympathetic response: (a) the widespread distribution of postganglionic fibres from a single sympathetic white ramus; (b) the localized distribution of postganglionic parasympathetic fibres.

The *IIIrd (oculomotor)* nerve outflow is relayed in the ciliary ganglion (see page 261), and provides the innervation of the sphincter pupillae and the ciliary muscle. Its stimulation produces constriction of the pupil and accommodation of the lens.

The *VIIth (facial)* outflow is relayed in the pterygopalatine ganglion (see page 264) and the submandibular ganglion (see page 273), which are responsible for the secretomotor supply of the lacrimal gland and of the submandibular and sublingual salivary glands, respectively.

The *IXth (glossopharyngeal)* outflow is relayed in the otic ganglion (see page 274) and is secretomotor to the parotid gland.

The *Xth (vagal)* outflow (see page 284) conveys by far the most important and largest contributions of the parasympathetic system. It is responsible for all the functions of the cranial outflow of this system enumerated above apart from the innervation of the eye and the secretomotor supply to the salivary and lacrimal glands. The efferent fibres are derived from the dorsal nucleus of X, which lies in the central grey matter of the lower medulla, and are distributed widely in the cardiac, pulmonary and alimentary plexuses already described. Postganglionic fibres are relayed from tiny ganglia that lie in the walls of the viscera concerned; in the gut these constitute the submucosal plexus of Meissner and the myenteric plexus of Auerbach.

The sacral outflow

The anterior primary rami of S2 and S3, and occasionally S4, give off nerve fibres termed the *pelvic splanchnic nerves* or nervi erigentes, which join the sympathetic pelvic plexuses for distribution to the pelvic organs. Tiny ganglia in the walls of the viscera then relay postganglionic fibres.

The sacral parasympathetic system has been aptly termed the 'mechanism for emptying'. It supplies visceromotor fibres to the muscles of the rectum (and perhaps the lower colon) and inhibitor fibres to the internal (involuntary) anal sphincter, motor fibres to the bladder wall and inhibitor fibres to the internal vesical sphincter. In addition, vasodilator fibres supply the erectile cavernous sinuses of the penis and clitoris.

Afferent parasympathetic fibres

Visceral afferent fibres from the heart, lung and the alimentary tract are conveyed in the vagus nerve, pass to the ganglion cells in the ganglion nodosum and thence pass to the dorsal nucleus of the vagus. Sacral afferents are conveyed in the pelvic splanchnic nerves S2, S3 and S4 and are responsible for visceral pain experienced in the bladder, prostate, rectum and uterus. The reference of pain from these structures to the sacral area, buttocks and posterior aspect of the thighs is explained by the similar segmental supply to the sacral dermatomes.

(Note that, although afferent fibres are conveyed in both sympathetic and parasympathetic nerves, they are completely independent of the autonomic system; they do not relay in the autonomic ganglia and terminate, just like somatic sensory fibres, in the dorsal ganglia of the spinal and cranial nerves. They merely use the autonomic nerves as a convenient anatomical conveyor system from the periphery to the brain (see page 333).

Part 6
The Cranial Nerves

Introduction

The 12 pairs of cranial nerves (Fig. 179) may be defined as those peripheral nerves that emerge from the brain; the remaining peripheral nerves leave the spinal cord and constitute the spinal nerves.

The first two nerves are atypical: I (the olfactory nerve) is formed by the unmyelinated central processes of the olfactory sensory cells; II (the optic nerve) represents a tract drawn out from the brain during the embryological development of the eye.

The remaining 10 pairs of nerves have a general plan of organization and it will simplify a complex subject if this is first considered.

The basic plan of the cranial nuclei

The nuclei of the 'true' cranial nerves are situated in the pons and medulla. Just as the spinal cord is organized to receive afferent fibres into its posterior grey columns and to relay postsynaptic efferent fibres from cells in the anterior grey columns, so these cranial nuclei are arranged into posterior afferent and anterior efferent cell groups. Here, some embryology must be interposed; the primitive tubular hindbrain resembles the spinal cord in cross-section, in being divided into a dorsal (alar) lamina and a ventral (basal) lamina, separated by the sulcus limitans, which defines the boundary between the dorsal afferent and ventral efferent components (Fig. 180a). In the region of the future pons, the hindbrain becomes kinked (the pontine flexure), so that its roof becomes stretched and its cavity flattened to form the 4th ventricle. However, the alar and basal laminae can still be identified and maintain their functional integrity.

In the basal lamina, three discontinuous columns of motor cells develop (Fig. 180b):

1 The *somatic efferent column* is the most ventrally placed and is equivalent to the anterior horn cells of the spinal cord. It is concerned with the innervation of those muscles of the head that are of myotomic origin (the extrinsic eye muscles and the muscles of the tongue), and is represented by the motor nuclei of III, IV, VI and XII.

2 The *branchial efferent column* is placed rather more dorsally and has no equivalent in the spinal cord. It is responsible for the innervation of the muscles derived from the branchial arches; the motor nucleus of V supplies the 1st arch, VII supplies the 2nd arch, IX supplies the 3rd arch and the nucleus ambiguus of X supplies the 4th and 6th arches.

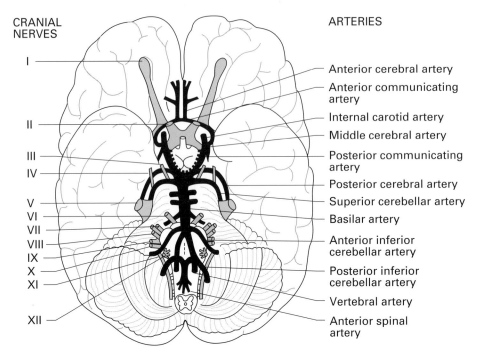

CRANIAL
NERVES

ARTERIES

I — Anterior cerebral artery

— Anterior communicating artery

— Internal carotid artery

II — Middle cerebral artery

III — Posterior communicating artery

IV

V — Posterior cerebral artery

VI — Superior cerebellar artery

VII — Basilar artery

VIII — Anterior inferior cerebellar artery

IX

X — Posterior inferior cerebellar artery

XI — Vertebral artery

XII — Anterior spinal artery

Fig. 179 The base of the brain, showing the relationships of the cranial nerve roots to the circle of Willis and its branches.

3 The *general visceral efferent column* is the most dorsal of the three. It is comparable to the lateral grey column of the spinal cord and, like it, is concerned with visceral autonomic innervation. It is represented by the Edinger–Westphal nucleus of III, the superior salivary nucleus of VII, the inferior salivary nucleus of IX and the dorsal motor nucleus of X, i.e. by those nuclei that are responsible for the cranial part of the parasympathetic outflow.

In the alar lamina, four cell groups that receive afferent fibres can be distinguished (Fig. 180b):

1 The *special somatic afferent column* (or audito-lateral column) is placed most dorsally. It is the relay station for the two special receptor organs of the ear – the cochlea and the vestibular apparatus.

2 The *general somatic efferent column* is next in line; it is concerned with innervation of the face and comprises the sensory nuclei of V.

3 The *special visceral afferent column* is concerned with the reception of taste impulses; it is represented by the nucleus of the tractus solitarius in the central grey matter of the medulla, which receives the chorda tympani fibres of VII and the gustatory fibres of IX and X.

4 The *general visceral afferent column* is placed nearest the equator of the medulla, receives afferent fibres from the viscera and is represented by the sensory component of the dorsal nucleus of the vagus.

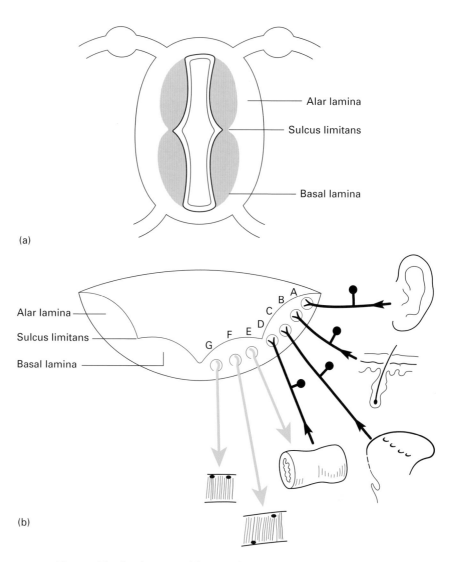

Fig. 180 The development of the cranial nerve nuclei. (a) The primitive hindbrain in transverse section. (b) The formation of the cranial nerve centres: A, special somatic afferent (auditory and vestibular); B, general somatic afferent (sensory V); C, special visceral afferent (taste VII, IX, X); D, general visceral afferent (afferent visceral X); E, general visceral efferent (efferent visceral III, VII, IX and X); F, branchial efferent (branchial arch muscles V, VII, IX and X); G, somatic efferent (cranial myotomes III, IV, VI and XII).

The olfactory nerve (I)

The fibres of the olfactory nerve, unlike other visceral afferent fibres, are the central processes of the olfactory cells and not the peripheral processes of a central group of ganglion cells (Fig. 181).

(a)

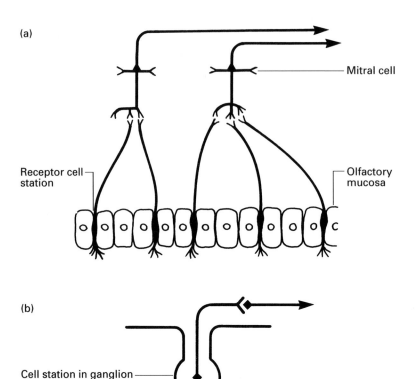

Mitral cell

Receptor cell
station

Olfactory
mucosa

(b)

Cell station in ganglion

Receptor

Fig. 181 The contrast between (a) the olfactory nerve and (b) a typical cranial sensory nerve. The cell station of the olfactory nerve is the receptor cell in the olfactory mucous membrane; the cell station of the typical sensory cranial nerve lies in its ganglion.

The central processes of the olfactory receptors pass upwards from the olfactory mucosa in the upper part of the superior nasal concha and septum in approximately 20 small nerve bundles, through the cribriform plate of the ethmoid bone to terminate by entering into synapse with the dendrites of mitral cells in the olfactory bulb. The mitral cells in turn send their axons back in the olfactory tract to terminate in the cortex of the uncus and the region of the anterior perforated space. The further course of the olfactory pathway is unknown, but it is now clear that the hippocampus–fornix system is not directly concerned with olfaction.

The sense of smell is not highly developed in humans and is easily disturbed by conditions affecting the nasal mucosa generally (e.g. the common cold). However, unilateral anosmia may be an important sign in the diagnosis of frontal lobe tumours. Tumours in the region of the uncus may give rise to the so-called uncinate type of fit, characterized by olfactory hallucinations associated with impairment of consciousness and involuntary chewing movements. Bilateral anosmia due to interruption of the 1st nerve is common after head injuries, particularly in association with anterior cranial fossa fractures.

As the bundles of the olfactory nerve pierce the cribriform plate, they receive a sheath of the meninges that blends with the extracranial neurilemma; this constitutes an important pathway of infection from the nasal cavity to the subarachnoid space.

The optic nerve (II)

The optic nerve is the nerve of vision. It is not a true cranial nerve but should be thought of as a brain tract that has been drawn out from the cerebrum; embryologically it is developed, together with the retina, as a lateral diverticulum of the forebrain. Devoid of neurilemmal sheaths, its fibres, like other brain tissues, are incapable of regeneration after division. Extension of the meninges and the subarachnoid space invest the nerve and fuse with the connective tissues of the sclera. Raised cerebrospinal fluid pressure is transmitted along this subarachnoid space extension and may compress the venous drainage of the optic nerve, thus producing papilloedema.

From a functional point of view, the retina can be regarded as consisting of three cellular layers: a layer of receptor cells – the rods and cones; an intermediate layer of bipolar cells; and a layer of ganglion cells (Fig. 182). From all parts of the retina, the axons of these ganglion cells converge on the optic disc, where they pierce the sclera to form the optic nerve.

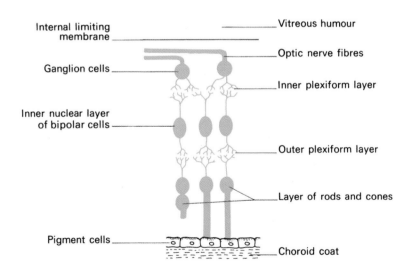

Internal limiting membrane — Vitreous humour
Optic nerve fibres
Ganglion cells — Inner plexiform layer
Inner nuclear layer of bipolar cells — Outer plexiform layer
Layer of rods and cones
Pigment cells — Choroid coat

Fig. 182 The composition of the retina.

The optic nerve passes backwards and medially to the optic foramen, through which it reaches the optic groove on the dorsum of the body of the sphenoid; it thus has an intra-orbital and an intracranial component. The intra-orbital part of its course is approximately 2.5 cm long. The nerve lies here in the orbital fat surrounded by the recti muscles; the ciliary ganglion lies lateral to it. Halfway along its intra-orbital path, the nerve is pierced inferomedially by the central artery and vein of the retina, which then pass forwards in its centre. In the optic foramen, the nerve lies superomedially to the ophthalmic artery (Fig. 183). Intracranially, the nerve is 1.25 cm long; it passes medially to the internal carotid artery to reach the optic chiasma. Here, all the fibres from the medial half of the retina, i.e. those concerned with the temporal visual field, cross over to the optic tract of the opposite side, whereas the fibres from the lateral half of the retina (nasal visual field) pass back in the optic tract of the same side (Fig. 184).

The great majority of the fibres in the optic tract end in the six-layered lateral geniculate body of the thalamus, but a small proportion, subserving pupillary and ocular reflexes, bypass the geniculate body to reach the superior colliculus. From the lateral geniculate body, the fibres of the optic radiation sweep laterally and backwards to the occipital visual cortex (the striate area above and below the calcarine fissure), where they terminate in such a way that the upper and lower halves of the retina are represented on the upper and lower lips of the fissure, respectively. It is interesting that the central part of the retina has a far greater cortical representation than the peripheral retina; a fact that correlates closely with the more intense visual acuity of the macular region.

Lesions of the retina or optic nerve result in unilateral blindness in the affected segment, but lesions of the optic tract and central parts of the visual pathway result in homonymous defects. Similarly, lesions of the optic chiasma, e.g. from an expanding pituitary tumour, will give rise to a bitemporal hemianopia, i.e. there will be a loss of vision in both temporal eye-fields.

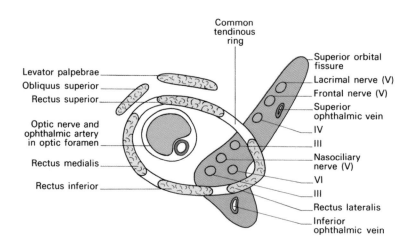

Fig. 183 The superior orbital fissure and the tendinous ring of origin of the extrinsic muscles, to show the relationships of the cranial nerves as they enter the orbit.

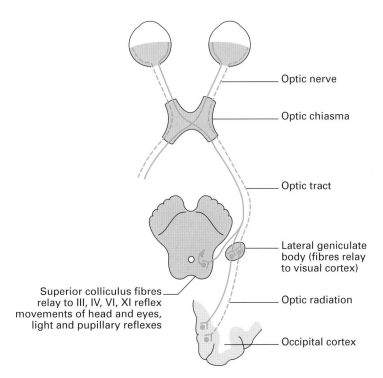

Optic nerve

Optic chiasma

Optic tract

Lateral geniculate body (fibres relay to visual cortex)

Superior colliculus fibres relay to III, IV, VI, XI reflex movements of head and eyes, light and pupillary reflexes

Optic radiation

Occipital cortex

Fig. 184 Plan of the optic pathway.

The oculomotor nerve (III)

In addition to supplying all the extrinsic eye muscles apart from the lateral rectus and superior oblique, the oculomotor nerve conveys the preganglionic parasympathetic fibres for the sphincter of the pupil and for the ciliary muscle. Its nucleus of origin lies in the floor of the cerebral aqueduct at the level of the superior colliculus and consists essentially of two components: the somatic efferent nucleus, which supplies the ocular muscles, and the nearby Edinger–Westphal nucleus of the general visceral efferent column from which the parasympathetic fibres are derived (Fig. 185).

The somatic efferent nuclei of the three nerves that control the movements of the eye, III, IV and VI, have exactly the same central connections. This, of course, one would expect since they function *en masse*. Voluntary eye movements are controlled from the opposite motor cortex, whose fibre connections descend in the contralateral pyramidal tract. Reflex eye movements depend on impulses received from the visual cortex and the vestibular apparatus; in consequence, these three cranial nerve nuclei are linked to the occipital cortex via the superior colliculus and the tectobulbar tract, and to the vestibular part of VIII via the medial longitudinal bundle. In addition, the Edinger–Westphal nucleus of III receives fibres from the optic nerve via the superior colliculus that subserve the light reflex (Fig. 184).

From the two nuclei of origin of III, fibres pass vertically through the mid-brain tegmentum to emerge just medial to the cerebral peduncle.

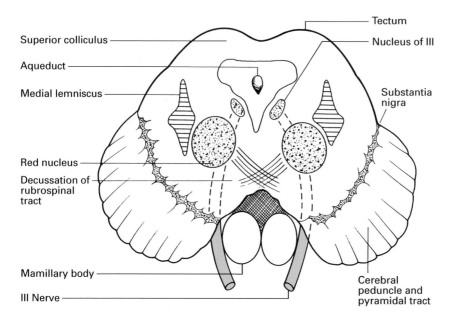

Superior colliculus

Aqueduct

Medial lemniscus

Red nucleus

Decussation of rubrospinal tract

Mamillary body

III Nerve

Tectum

Nucleus of III

Substantia nigra

Cerebral peduncle and pyramidal tract

Fig. 185 Section of the midbrain at the level of the red nucleus to show the IIIrd nerve nucleus.

Passing forwards between the superior cerebellar and posterior cerebral arteries (Fig. 179), the nerve pierces the dura mater to run in the lateral wall of the cavernous sinus (Fig. 186) as far as the superior orbital fissure. Before entering the fissure, it divides into a superior and inferior branch; both branches enter the orbit through the tendinous ring, from which the recti arise (Fig. 183). The superior branch passes lateral to the optic nerve to supply the superior rectus muscle and levator palpebrae superioris; the inferior branch supplies three muscles – the medial rectus, the inferior rectus and the inferior oblique (the nerve to the last conveying the parasympathetic fibres to the ciliary ganglion) (see page 261).

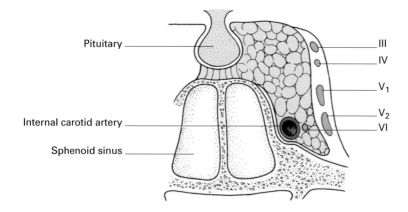

Pituitary

Internal carotid artery

Sphenoid sinus

III

IV

V_1

V_2

VI

Fig. 186 The cavernous sinus to show the relationships of cranial nerves III, IV, V and VI.

Complete division of III results in a characteristic group of signs:

1 ptosis, due to paralysis of the levator palpebrae superioris;
2 a divergent squint, due to the unopposed action of the superior oblique and lateral rectus muscles, which rotate the eyeball laterally;
3 dilatation of the pupil, the dilator action of the sympathetic fibres being unopposed;
4 loss of the accommodation–convergence and light reflexes, due to ciliary muscle paralysis;
5 double vision.

The trochlear nerve (IV)

The trochlear nerve is the slenderest of the cranial nerves and supplies only one eye muscle, the superior oblique. Its somatic efferent nucleus of origin lies in a similar position to that of the IIIrd nerve at the level of the inferior colliculus immediately beneath the aqueduct. The central connections of this nucleus are described with III. From this nucleus, fibres pass dorsally around the cerebral aqueduct and decussate in the superior medullary velum (Fig. 187).

Emerging immediately behind the inferior colliculus, the nerve winds round the cerebral peduncle and then passes forwards between the superior cerebellar and posterior cerebral arteries to pierce the dura. It then runs forwards in the lateral wall of the cavernous sinus (Fig. 186), between the oculomotor and ophthalmic nerves, to enter the orbit through the superior

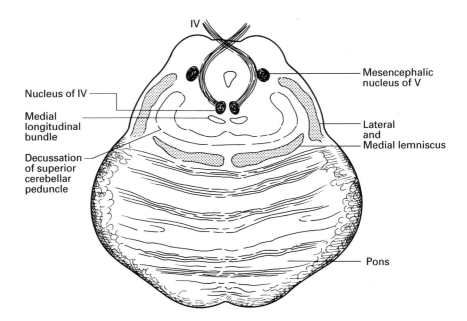

Fig. 187 Section through the upper pons to show the nucleus of the IVth nerve.

orbital fissure, lateral to the tendinous ring from which the recti take origin (Fig. 183). It then passes medially over the optic nerve to enter the superior oblique muscle.

A lesion of the trochlear nerve results in paralysis of the superior oblique muscle with the result that diplopia occurs when the patient attempts to turn the eye downwards and laterally.

The trigeminal nerve (V)

The trigeminal is the largest of the cranial nerves. Because it is the principal sensory nerve of the face, orbit, nose and mouth, and because its branches are eminently suitable for accurate anaesthetic blockade, no excuse need be offered to the reader for the detail with which this nerve is to be considered.

The trigeminal nerve has a large sensory and small motor root; in addition, it is associated with four autonomic ganglia. A summary of its distribution is as follows:

1 *Sensory* – to the face and the scalp back as far as the vertex; the mucosa of the nasal cavity, accessory nasal sinuses and much of the nasopharynx; the orbit and eyeball; the mucosa of the mouth, gums and palate; the anterior two-thirds of the tongue and the teeth.
2 *Motor* – to the muscles of mastication, mylohyoid, the anterior belly of digastric, tensor palati and tensor tympani.
3 *Ganglionic connections* – to the ciliary, pterygopalatine, submandibular and otic ganglia.

The *motor nucleus* of the trigeminal nerve, which belongs to the branchial efferent column, is situated in the upper pons, immediately below the lateral part of the floor of the 4th ventricle (Figs 188, 189). It receives corticobulbar fibres from both sides of the cerebral motor cortex, particularly the contralateral side.

The *sensory nucleus*, which represents part of the general somatic afferent column, is in three parts (Figs 188, 189). Sensory fibres from the trigeminal ganglion on entering the pons divide into ascending and descending tracts; the ascending fibres pass to the *mesencephalic nucleus of the trigeminal nerve* in the central grey matter of the mid-brain, and to the *superior* (or *principal*) *sensory nucleus* that lies on the lateral side of the motor nucleus, separating the latter from the superior cerebellar peduncle. The descending fibres constitute the spinal tract of the trigeminal nerve, which runs downwards through the whole length of the pons and medulla and blends inferiorly with the substantia gelatinosa, where afferent nerves synapse with the lateral reticular formation. These nerves cap the posterior horn of the spinal grey matter. Immediately deep to this tract is the *nucleus of the spinal tract*, which runs from the superior nucleus rostrally to the spinal cord caudally.

Each of these three sensory nuclei subserves different modalities. The mesencephalic nucleus is concerned with proprioception, the superior nucleus with touch and the spinal tract nucleus with pain and temperature. Within the nucleus of the spinal tract there is an orderly representation of the three divisions of the trigeminal nerve: the ophthalmic (V') fibres

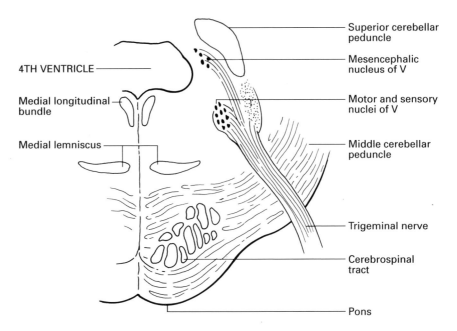

Fig. 188 The nuclei of the trigeminal nerve in a transverse section of the pons.

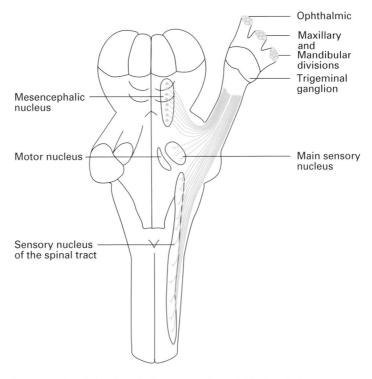

Fig. 189 Plan of the trigeminal nerve and its nuclei in dorsal view.

terminate caudally, then the maxillary (V'') and finally the mandibular (V''') most rostrally. Moreover, the fibres are distributed in echelon, so that V''' is represented dorsally, then V'' and finally V' most ventrally.

The central fibres that relay from the three sensory nuclei of V decussate, then ascend in the trigeminal lemniscus to the lateral nucleus of the thalamus, from which further relay occurs to the face area of the postcentral (sensory) cerebral gyrus.

The two roots of the nerve emerge from the ventrolateral aspect of the pons near its upper border (Fig. 179); the larger, lateral root (portio major) is sensory, the smaller, medial root (portio minor) is motor. The nerve passes ventrally through the cisterna pontis, and has a course of approximately 1 cm before the sensory root swells into the trigeminal ganglion, which is the first cell station for its sensory fibres.

The *trigeminal ganglion*, which is also termed the semilunar or Gasserian ganglion, is equivalent to the dorsal sensory ganglion of a spinal nerve. It is crescent-shaped, with its convex surface pointing laterally, and is situated within an invaginated pocket of dura immediately inferior to the anterior attachment of the tentorium cerebelli. The ganglion lies near the apex of the petrous temporal bone, which is somewhat hollowed for it, and overlaps onto the cartilage that fills the foramen lacerum. The motor root of the trigeminal nerve and the greater petrosal nerve both pass deep to the ganglion. Above is the hippocampal gyrus of the temporal lobe of the cerebrum; medially lie the internal carotid artery and the posterior part of the cavernous sinus.

From the concave medial aspect of the ganglion, fibres pass backwards below the superior petrosal sinus within the apex of the tentorium cerebelli to reach the pons. From the antero-inferior aspect of the ganglion pass the 1st (ophthalmic), 2nd (maxillary) and 3rd (mandibular) divisions of the nerve, each of which supplies an embryologically distinct facial segment; the frontonasal, maxillary and mandibular processes of fetal development, respectively (Fig. 190). V' itself splits up into three branches that pass

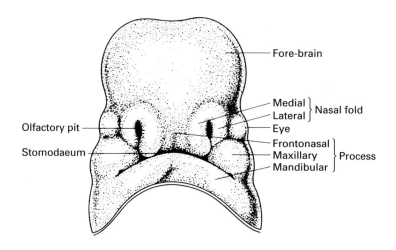

Fore-brain

Medial ⎫
Lateral ⎬ Nasal fold

Olfactory pit

Eye

Frontonasal ⎫
Maxillary ⎬ Process
Mandibular ⎭

Stomodaeum

Fig. 190 The development of the face: the frontonasal, maxillary and mandibular processes of the fetus are, respectively, innervated by the 1st, 2nd and 3rd divisions of the trigeminal nerve.

forwards and somewhat upwards to the superior orbital fissure, V″ passes anteriorly through the foramen rotundum into the pterygopalatine fossa and V‴ dives almost vertically downwards to escape through the foramen ovale into the infratemporal fossa.

CLINICAL NOTE

Trigeminal ganglion block (Fig. 191)

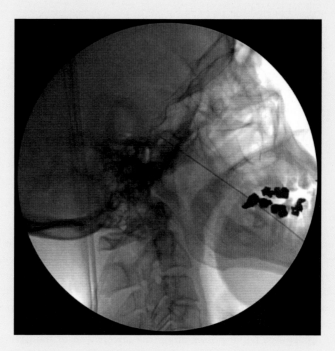

Fig. 191 Screening radiographic image of a needle passing through the foramen ovale for trigeminal nerve block.

The most common indication for this trigeminal ganglion block is intractable trigeminal neuralgia. However, it has been used for intractable cancer pain and cluster headache. It is believed that in trigeminal neuralgia the nerve is compressed proximally close to the brainstem by a tortuous or ectasic blood vessel, which causes demyelination.

Microvascular decompression (MVD), surgical relief of the compressed nerve, produces the most sustained response but requires a suboccipital craniotomy. The most common finding is compression of the nerve at the entry zone by a branch of the superior cerebellar artery. Following surgery 4% of patients have major problems such as cerebrospinal fluid leakage (<2%) or cerebellar infarction (0.45%).

When decompression is not indicated or when MVD is not available or not tolerable, other neuroablative/destructive procedures have been used to target the Gasserian ganglion. Those in common usage are radiofrequency gangliolysis, glycerol injection into Meckel's cave and balloon compression. In addition, stereotactic radiosurgery (gamma knife), in which a focused beam of radiation is aimed at the trigeminal root in the posterior fossa, is available in some centres.

The percutaneous procedures are performed under radiographic control using an image intensifier or computed tomography guidance. The introducer/needle is inserted below the posterior one-third of the zygomatic bone, opposite the second upper molar tooth. It is then passed posteromedially behind the pterygoid plate to the foramen ovale, which is identified radiologically. The needle is advanced 1 cm beyond the foramen into the ganglion, and here paraesthesia is usually experienced in the trigeminal distribution. At this point, aspiration is performed to ensure that the dura has not been punctured, and the non-penetration of the thecal coverings by the needle is confirmed by injection of radio-opaque dye. If this contrast medium remains localized in the region of the foramen, then injection compression or radiofrequency ablation may take place.

Approximately 4% of patients develop some degree of corneal hypoaesthesia and anaesthesia dolorosa with these techniques. Some sensory loss occurs in 50%.

The ophthalmic nerve (V′)

The ophthalmic nerve (Figs 192, 193), the first division of the trigeminal nerve, is entirely sensory. It is distributed to the eyeball, together with its overlying conjunctiva, protective upper lid and adjacent lacrimal gland; to the skin of the forehead, nose and scalp back as far as the vertex; to the mucous membrane of the medial and lateral walls of the anterior part of the nose; and to the adjacent frontal and ethmoidal sinuses.

The ophthalmic nerve passes along the lateral wall of the cavernous sinus, below cranial nerves III and IV (Fig. 186), to reach the superior orbital fissure, where it divides into its lacrimal, frontal and nasociliary branches.

Because of the proximity of the ophthalmic division of V to the optic nerve, a block of the ophthalmic nerve by local anaesthetic agents, which is occasionally indicated in eye surgery, may cause temporary interference with vision. Epinephrine should not be used in local anaesthetic solutions employed for this purpose because the central artery of the retina is an end-artery, and the nerve should never be blocked with neurolytic agents because of the risk of producing permanent blindness.

The *lacrimal nerve* is the smallest of the three branches. It passes into the orbit through the lateral part of the superior orbital fissure above the fibrous ring of origin of the extrinsic orbital muscles (Fig. 183). It supplies a branch to the lacrimal gland, which transmits parasympathetic secretomotor fibres from the pterygopalatine ganglion; these fibres reach the lacrimal nerve from a communicating twig derived from the zygomatic branch of the maxillary division. The lacrimal nerve emerges from the orbit below the lateral extremity of the orbital margin; here, its terminal fibres supply the conjunctiva and a patch of skin of the upper eyelid adjacent to the outer canthus.

The *frontal nerve* is the largest branch of the ophthalmic nerve. It passes through the superior orbital fissure above the orbital ring (Fig. 183) and above levator palpebrae superioris. Within the orbit, it divides into the (larger) *supra-orbital* and the (smaller) *supratrochlear* branch (Fig. 192).

The supra-orbital nerve ascends through the notch (or occasionally the foramen) in the supra-orbital ridge, supplies branches to the medial side of the upper eyelid then supplies the forehead and scalp backwards to the vertex.

The supratrochlear nerve passes above the pulley of the superior oblique muscle to supply the conjunctiva and the skin of the upper eyelid near the inner canthus, the medial part of the forehead just above the orbit and the root of the nose.

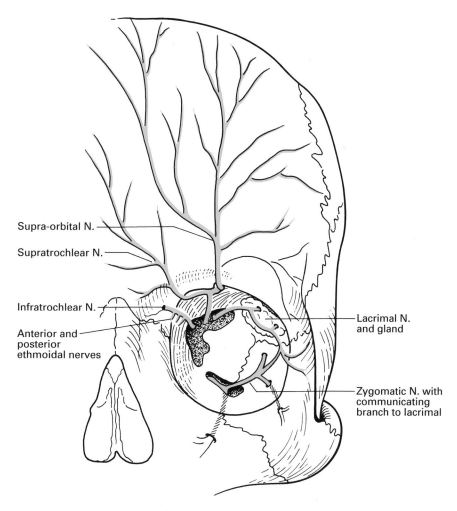

Fig. 192 Frontal view of the orbit to show the branches of the ophthalmic division of V.

If three fingers are placed side by side from the midline of the forehead outwards, the second finger lies on the orbit where this is crossed by the supratrochlear nerve, and the third finger lies over the supra-orbital nerve. The latter point may often be identified by palpation of the supra-orbital notch, pressure on which is unpleasantly painful.

The *nasociliary nerve* enters the superior orbital fissure within the tendinous ring of origin of the recti (Fig. 183), crosses obliquely to the medial wall of the orbit, passing as it does so medially and above the optic nerve, and then enters the anterior ethmoidal foramen. Here, it becomes the *anterior ethmoidal nerve*, which runs along the anterior cranial fossa on the cribriform plate to enter the nasal cavity, through an aperture near the crista galli, where it splits into septal and lateral branches. The septal branch supplies the mucosa over the anterior part of the nasal septum; the lateral branch supplies the anterior part of the lateral wall, then emerges as the *external nasal nerve* between the lower border of the nasal bone and the lateral nasal cartilage to supply the skin over the ala and the tip of the nose (the nasociliary nerve thus supplies the cartilaginous tip of the nose on both its inner and outer aspects).

In its course, the nasociliary nerve gives off the following branches (Fig. 193):

1 a sensory contribution to the ciliary ganglion (see below);
2 two *long ciliary nerves*, which enter the back of the eyeball; they are sensory but also transmit sympathetic dilator pupillae fibres;
3 the *posterior ethmoidal nerve*, which is given off as the nasociliary nerve reaches the medial wall of the orbit, and which supplies branches to the posterior ethmoidal air cells via the posterior ethmoidal foramen;

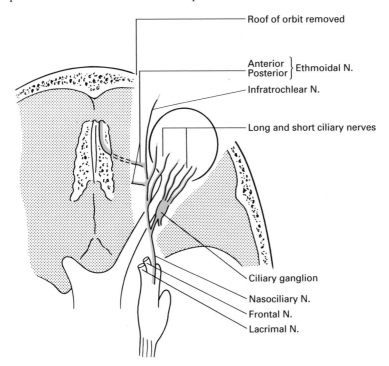

Roof of orbit removed

Anterior } Ethmoidal N.
Posterior }

Infratrochlear N.

Long and short ciliary nerves

Ciliary ganglion

Nasociliary N.

Frontal N.

Lacrimal N.

Fig. 193 Superior view of the orbit to show the ophthalmic division of V and the distribution of the nasociliary nerve.

4 the *infratrochlear nerve*, which is given off immediately before the nasocil-iary nerve enters the anterior ethmoidal foramen; it leaves the orbit, as its name implies, below the trochlea and innervates the skin of the side of the nose and the conjunctiva near the inner canthus.

Ciliary ganglion

The *ciliary ganglion* (Fig. 193) lies near the apex of the orbit between the optic nerve and the rectus lateralis. It is a tiny structure, approximately 1 mm in diameter, which receives parasympathetic, sympathetic and sen-sory fibres.

Its *parasympathetic component* is derived from the oculomotor nerve (III) via its branch to the inferior oblique muscle, these fibres originating in the Edinger–Westphal nucleus (see page 251). Relay takes place in the ciliary ganglion, and postganglionic fibres then pass in the short ciliary nerves, about six in number, to the eyeball, where they supply the sphincter pupil-lae and the ciliary muscle. Stimulation results in pupillary constriction and in accommodation of the lens.

Sympathetic fibres reach the ganglion from the superior cervical ganglion via the internal carotid plexus.

Sensory fibres are derived from the nasociliary branch of the ophthalmic nerve. Both the sympathetic and sensory components pass through the cil-iary ganglion without synapse to reach the eye via the short ciliary nerves, where they are, respectively, vasoconstrictor and sensory to the globe of the eye. (Note that the majority of sympathetic dilator pupillae nerve fibres are transmitted to the eye in the long ciliary branches of the nasociliary nerve.)

The maxillary nerve (V″)

The maxillary nerve (Figs 194, 195, 196, 197), the second division of the trigeminal nerve, is intermediate in size as well as position between the first and third divisions; it is entirely sensory. Its course is through four anatomical zones: skull base, pterygopalatine fossa, infra-orbital canal and, finally, the subcutaneous tissues of the cheek.

The maxillary nerve arises from the anterior border of the trigeminal ganglion, and then runs along the lower lateral wall of the cavernous sinus below the ophthalmic nerve (Fig. 186). It then leaves the skull base through the foramen rotundum and traverses the pterygopalatine fossa. The nerve is now named the *infra-orbital nerve* and lies, in turn, in the infra-orbital groove and then the infra-orbital canal of the orbital aspect of the maxilla. The nerve emerges from the infra-orbital foramen, where it lies beneath levator labii superioris, then divides into branches that are distributed to the lower eyelid, the side of the nose, the cheek and the upper lip.

The branches of the maxillary nerve are conveniently divided into four groups from the four successive stages of the nerve's anatomical pathway:
1 intracranial – meningeal branch;
2 pterygopalatine fossa – zygomatic, ganglionic and posterior superior alveolar nerves;

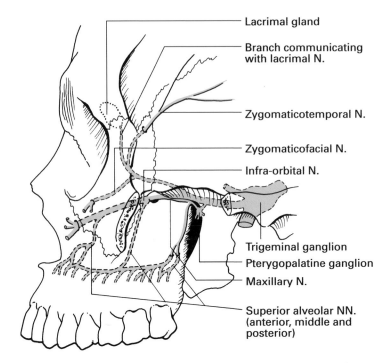

Lacrimal gland

Branch communicating
with lacrimal N.

Zygomaticotemporal N.

Zygomaticofacial N.

Infra-orbital N.

Trigeminal ganglion

Pterygopalatine ganglion

Maxillary N.

Superior alveolar NN.
(anterior, middle and
posterior)

Fig. 194 Distribution of the maxillary division of V excluding the branches of the
pterygopalatine ganglion; lateral view.

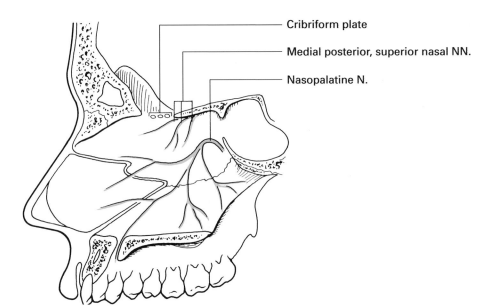

Cribriform plate

Medial posterior, superior nasal NN.

Nasopalatine N.

Fig. 195 Branches of the maxillary division associated with the pterygopalatine
ganglion on the medial wall of the nose.

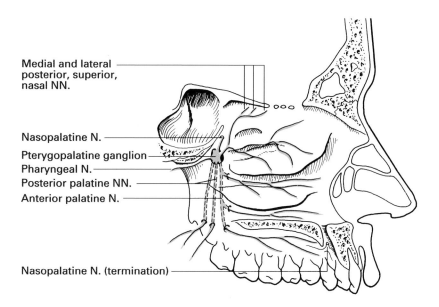

Medial and lateral posterior, superior, nasal NN.

Nasopalatine N.

Pterygopalatine ganglion

Pharyngeal N.

Posterior palatine NN.

Anterior palatine N.

Nasopalatine N. (termination)

Fig. 196 Branches of the maxillary division associated with the pterygopalatine ganglion on the lateral wall of the nose and the palate.

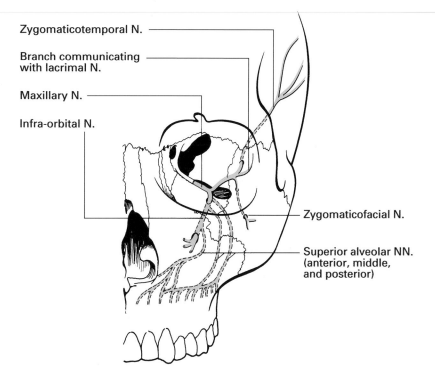

Zygomaticotemporal N.

Branch communicating with lacrimal N.

Maxillary N.

Infra-orbital N.

Zygomaticofacial N.

Superior alveolar NN. (anterior, middle, and posterior)

Fig. 197 Anterior view of the branches of the maxillary division of V excluding the branches of the pterygopalatine ganglion.

3 infra-orbital canal – middle superior alveolar and anterior superior alveolar nerves;
4 face – palpebral, nasal and labial branches.

The *meningeal branch* supplies the dura of the middle cranial fossa.

The *zygomatic nerve* arises from the maxillary nerve as this traverses the pterygopalatine fossa. It passes through the inferior orbital fissure to run on the lateral wall of the orbit, where it divides into two branches:

1 The *zygomaticotemporal nerve* traverses the zygomaticotemporal canal in the zygomatic bone to enter the temporal fossa; thence, it ascends to supply the skin of the temporal region. While still within the orbit, this nerve gives a twig to the lacrimal nerve, along which parasympathetic secretomotor fibres from the pterygopalatine ganglion are conveyed to the lacrimal gland.
2 The *zygomaticofacial nerve* pierces the zygomaticofacial foramen of the zygoma, which transmits it from the orbit onto the skin over the cheek prominence.

The *ganglionic branches* (two in number) comprise the sensory roots of the pterygopalatine ganglion (see below).

The *posterior superior alveolar nerve*, which may be double, descends over the posterior surface of the maxilla, then enters the posterior dental canal (which again may be double) on the posterior aspect of the maxilla, and gives branches to each molar tooth. In addition, branches are given off to the mucosa of the maxillary sinus.

The *middle superior alveolar nerve* originates in the posterior part of the infra-orbital canal, then runs downwards in the lateral wall of the maxilla to supply the two upper premolars.

The *anterior superior alveolar nerve* arises at the anterior end of the infra-orbital canal and descends in the anterior maxillary wall to innervate the upper canine and incisor teeth. A tiny branch pierces the lateral wall of the inferior meatus to supply the mucous membrane of the lower anterior part of the lateral wall and the floor of the nasal cavity.

As the infra-orbital nerve emerges from the infra-orbital foramen, it breaks up into a spray of branches:

1 the *palpebral branches* supply the skin of the lower eyelid and the conjunctiva;
2 the *nasal branches* supply the skin of the side of the nose;
3 the *labial branches* supply the skin and mucous membrane of the upper lip and the anterior part of the cheek.

Pterygopalatine ganglion

The *pterygopalatine ganglion* (Fig. 196) is associated closely with the maxillary nerve and is deeply placed in the upper part of the pterygopalatine fossa. It receives parasympathetic, sympathetic and sensory nerve fibres.

The *parasympathetic component* is derived from the greater petrosal nerve, which originates from the geniculate ganglion of the facial nerve (VII). This nerve traverses the petrous temporal bone then runs in a groove on the anterior surface of the bone deep to the trigeminal ganglion to enter the foramen lacerum. Here, it is joined by the deep petrosal nerve to form the *nerve of the pterygoid canal* (the Vidian nerve), which passes through the pterygoid canal to reach the pterygopalatine ganglion. These parasympathetic fibres, having arrived at the ganglion, have not completed their complicated journey. They are transmitted via the zygomaticotemporal branch of the maxillary nerve to the lacrimal branch of the ophthalmic nerve, by which they arrive at their final destination as secretomotor fibres to the lacrimal gland.

Sympathetic fibres, derived from the internal carotid plexus, form the deep petrosal nerve that, as described above, reaches the ganglion via the nerve of the pterygoid canal.

The *sensory component* is derived from the two sphenopalatine branches of the maxillary nerve.

The sensory and sympathetic (vasoconstrictor) branches of the ganglion are distributed to the nose, nasopharynx, palate and orbit via the following branches (Figs 195, 196):

1 The *nasopalatine (long sphenopalatine) nerve* passes medially through the sphenopalatine foramen, crosses the roof of the nasal cavity, then passes downwards and forwards along the nasal septum, grooving the vomer as it does so, to reach the incisive foramen and thence the mucous membrane of the roof of the mouth (Fig. 195). It supplies filaments to the posterior part of the nasal roof, to the nasal septum and to those parts of the gums and anterior part of the hard palate that are in relation to the incisor teeth.

2 The *medial* and *lateral posterior superior nasal nerves* also pass medially through the sphenopalatine foramen; they supply sensory fibres to the superior and middle conchae and to the posterior part of the nasal septum (Fig. 196).

3 The *greater (anterior) palatine nerve* descends through the greater palatine canal, and then emerges onto the hard palate from the greater palatine foramen just posterior to the palatomaxillary suture (Fig. 198). It innervates the mucosa of the gums and hard palate as far forward as the level of the canine teeth. Other fibres pass backwards to serve both aspects of the soft palate, and nasal branches pierce openings in the perpendicular plate of the palatine bone to supply the region of the inferior nasal concha.

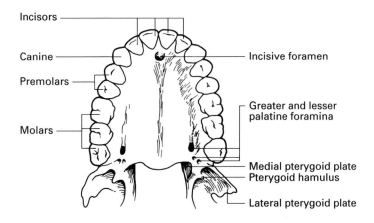

Fig. 198 The hard palate to show its foramina of exit.

Incisors

Canine — Incisive foramen

Premolars

Molars

Greater and lesser palatine foramina

Medial pterygoid plate
Pterygoid hamulus

Lateral pterygoid plate

4 The *lesser* (*middle and posterior*) *palatine nerves*, two or occasionally three in number, pass through the greater palatine canal in company with the greater palatine nerve, but emerge through separate lesser palatine foramina, which perforate the inferior and medial aspects of the tubercle of the palatine bone (Fig. 198). They supply the soft palate, uvula and tonsil.

5 The *pharyngeal nerve* passes backwards through the pharyngeal canal in the posterior wall of the pterygopalatine fossa to supply an area of nasopharyngeal mucosa immediately behind the orifice of the Eustachian tube.

6 The *orbital branches* are small; they constitute two or three fine twigs that pass through the superior orbital fissure to supply the adjacent periosteum, and perhaps also to carry some secretomotor fibres from the pterygopalatine ganglion to the lacrimal gland.

A digression on the pterygopalatine fossa

It might be wise here to consider briefly the anatomy of the pterygopalatine fossa, through whose openings pass the numerous branches of the second part of the maxillary nerve and of the pterygopalatine ganglion.

The fossa is an elongated, narrow, pyramidal interval below the apex of the orbit, lying between the upper part of the posterior surface of the maxilla in front and the greater wing and the root of the pterygoid process of the sphenoid behind. The roof is formed by the inferior surface of the body of the sphenoid medially, but laterally the roof is deficient and the fossa opens freely into the orbit via the posterior part of the superior orbital fissure. Medially, the space is closed by the vertical plate of the palatine bone, but laterally the fossa is wide open and communicates with the infratemporal fossa through the pterygomaxillary fissure. Inferiorly, the anterior and posterior walls come into apposition, thus sealing off the base of the fossa.

The entrances into and exits from the pterygopalatine fossa:

1 Medial – the sphenopalatine foramen is the gap in the upper end of the vertical plate of the palatine bone between its orbital and sphenoidal processes; it transmits into the nasal cavity, the nasopalatine and the medial and lateral posterior superior nasal nerves and the accompanying branches from the maxillary vessels.

2 Posterior:

  a the foramen rotundum admits the maxillary nerve into the fossa;

  b the pterygoid canal transmits the nerve of the pterygoid canal, which is formed by the fusion of the deep petrosal nerve and the greater superficial petrosal nerve (see page 264);

  c the pharyngeal canal conveys the pharyngeal branch of the pterygopalatine ganglion, together with its accompanying vessels, into the nasopharynx.

3 Anterior – the inferior orbital fissure leads from the upper end of the fossa forwards into the orbit. It transmits the maxillary nerve, zygomatic

nerve, orbital branches of the pterygopalatine ganglion and the infra-orbital vessels.

4 Lateral – the pterygomaxillary fissure leads into the infratemporal fossa. It is the inlet for the maxillary artery into the pterygopalatine fossa. The posterior superior alveolar branch of the maxillary nerve emerges through this fissure to enter the posterior dental canal on the posterior aspect of the maxilla.

5 Inferior – the greater palatine canal transmits the greater (anterior) and lesser (middle and posterior) palatine nerves and vessels that appear on the hard palate through the greater and lesser palatine foramina.

CLINICAL NOTE

Maxillary nerve block

Maxillary nerve block is performed for acute or chronic herpetic neuralgia, trigeminal neuralgia and most commonly for cancer pain. It provides for anaesthesia of the hemi-maxilla. Injection is performed as the nerve lies in the pterygopalatine fossa after emerging from the foramen rotundum. It is reached by inserting a needle through the mid-point of the coronoid notch beneath the zygomatic arch. The needle is advanced approximately 3.5–5.0 cm perpendicular to the base of the skull to reach the lateral pterygoid plate. The needle is withdrawn a little and advanced in an antero-superior direction to enter the pterygopalatine fossa. Paraesthesia is often produced. After aspiration to confirm that the needle is not lying within a blood vessel, 3–5 ml of local anaesthetic is injected. The presence of a plexus of veins in this area means that haematoma formation occasionally occurs (Fig. 199). This injection is most commonly performed using image-intensifier guidance.

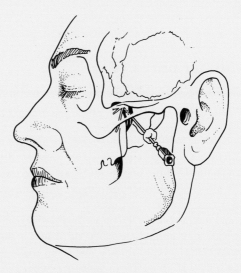

Fig. 199 The anatomy of maxillary nerve block.

It is occasionally useful to perform a localized nerve block of the infra-orbital nerve. The infra-orbital foramen can usually be palpated midway between the outer canthus of the eye and the alar process of the nose. The supra-orbital notch (or foramen), infra-orbital foramen and the mental foramen all lie in the same sagittal plane (Fig. 200).

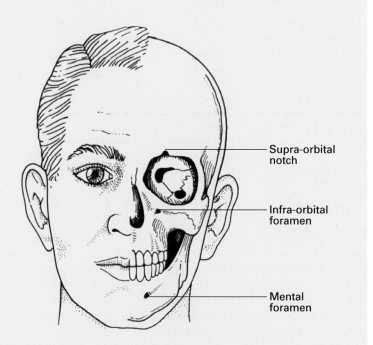

Supra-orbital notch

Infra-orbital foramen

Mental foramen

Fig. 200 The supra-orbital notch (or foramen), infra-orbital foramen and the mental foramen all lie in the same sagittal plane.

The mandibular nerve (V''')

The *mandibular nerve* (Figs 201, 202), the third division of the Vth nerve, is the largest, has the widest distribution and is the only one with a motor component.

It is the sensory nerve to the temporal region, the tragus and front of the helix, to the skin over the mandible and the lower lip, and to the mucosa of the anterior two-thirds of the tongue and floor of the mouth. Its motor fibres supply the muscles of mastication, tensor tympani, tensor palati, the mylohyoid and the anterior belly of digastric.

The sensory and motor roots of the nerve pass individually through the foramen ovale and unite immediately beyond it into a short trunk that lies deep to the lateral pterygoid muscle and upon the tensor palati, the latter separating it from the Eustachian (auditory) tube. The otic ganglion (see page 274) is situated immediately medial to the nerve and the middle meningeal artery is directly behind it.

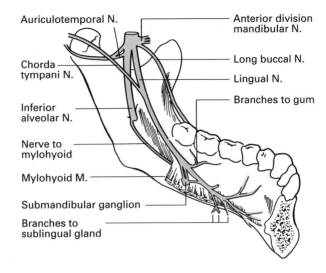

Fig. 201 The distribution of the lingual nerve.

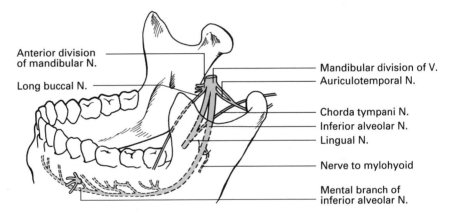

Fig. 202 The distribution of the inferior alveolar nerve.

CLINICAL NOTE

Mandibular nerve block

The technique is similar to that for a maxillary nerve block: a needle is inserted through the mid-point of the coronoid notch beneath the zygomatic arch. The needle is advanced approximately 3.5–5.0 cm perpendicular to the base of the skull to reach the lateral pterygoid plate. The needle is withdrawn and directed posterosuperiorly so that it moves off the posterior surface of the lateral pterygoid plate to meet the mandibular nerve as this emerges from the foramen ovale. At this point, the nerve lies in close relationship to the middle meningeal artery, the maxillary artery and the pterygoid plexus of veins. Successful nerve block will paralyse the muscles of mastication as well as producing anaesthesia of the lower jaw, the side of the tongue and the skin overlying the mandible (Fig. 203). Image intensification is commonly used when using this route.

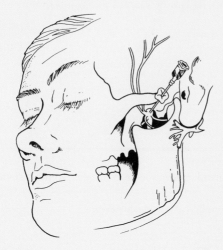

Fig. 203 The anatomy of mandibular nerve block.

Distribution of the mandibular nerve

The mandibular nerve soon divides into a smaller anterior and larger posterior trunk; the branches of the nerve and its trunk may be summarized thus:

1 Undivided trunk:
   a nervus spinosus (sensory);
   b nerve to medial pterygoid (motor).
2 Anterior trunk:
   a buccal nerve (sensory);
   b masseteric nerve (motor);
   c deep temporal nerves (motor);
   d nerve to lateral pterygoid (motor).
3 Posterior trunk:
   a auriculotemporal nerve (sensory);
   b lingual nerve (sensory);
   c inferior alveolar nerve (mixed).

(Together with the branches of distribution of the otic and submandibular ganglia.)

The *nervus spinosus* accompanies the middle meningeal artery into the skull through the foramen spinosum and, like the artery, it divides into anterior and posterior branches that supply the adjacent dura.

The *nerve to the medial pterygoid* supplies this muscle on its deep aspect. Motor fibres pass from the nerve to the otic ganglion, from which they are transmitted to tensor palati and tensor tympani.

The *buccal nerve* passes between the heads of the lateral pterygoid, runs downwards deep to temporalis and reaches the subcutaneous tissues of the cheek at the anterior margin of the ramus of the mandible. It supplies the skin over the anterior part of the cheek and also, via fibres that pierce buccinator, the mucous membrane of the inner aspect of the cheek and the lateral aspect of the gum adjacent to the molar teeth of the mandible.

The *masseteric nerve* appears above the upper border of the lateral pterygoid muscle, and passes laterally through the mandibular notch to the masseter. In addition, a twig of supply passes to the temporomandibular joint.

The *deep temporal nerves*, anterior, posterior and occasionally middle, pass above the upper border of the lateral pterygoid to the temporal muscle.

The *nerve to the lateral pterygoid* passes directly to this muscle.

The *auriculotemporal nerve* arises by two roots from the posterior aspect of the posterior trunk close to its origin. The two roots encircle the middle meningeal artery, join together, and then the common trunk passes backwards, first deep to the lateral pterygoid muscle, then deep to the neck of the mandible, where it lies between the bone and the sphenomandibular ligament. This ligament is a thin band that stretches from the spine of the sphenoid to the lingula immediately in front of the mandibular foramen; the auriculotemporal nerve is one of the many structures that pass between it and the mandible, the others being the lateral pterygoid muscle insertion, the maxillary vessels, the inferior alveolar vessels and nerve and a deep lobule of the parotid gland.

The auriculotemporal nerve emerges from behind the neck of the mandible just below the temporomandibular joint, where it lies deep to the parotid gland. The nerve ascends over the zygomatic arch in front of the ear and immediately behind the superficial temporal vessels; since the pulsations of the artery are here readily palpable, the precise surface marking of the nerve can be demarcated at this point.

The auriculotemporal nerve gives off the following branches:
1 auricular – to the skin of the tragus and the adjacent lateral aspect of the helix;
2 superficial temporal – this ramifies over the skin of the temporal region and the lateral aspect of the scalp;
3 branches to the external auditory meatus – (usually two), which supply the skin of the meatus and the tympanic membrane;
4 articular – to the temporomandibular joint;
5 parotid – which conveys secretomotor, sympathetic and sensory fibres to the salivary gland (see otic ganglion; page 274).

The *lingual nerve* commences between the tensor palati and lateral pterygoid muscles, receives the chorda tympani branch of the facial nerve (see Fig. 207), then, at the lower border of the lateral pterygoid, passes forwards between the ramus of the mandible and the medial pterygoid to reach the floor of the mouth by passing below the mandibular origin of the superior pharyngeal constrictor muscle. It then lies immediately under the mucosa of the gum on the inner surface of the mandible just below the level of the

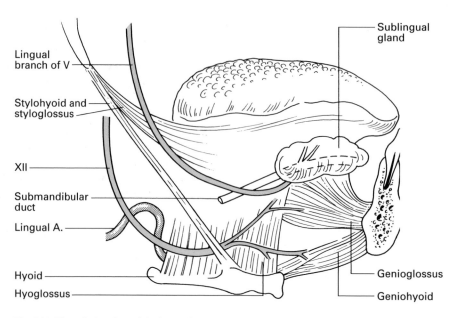

Lingual branch of V

Stylohyoid and styloglossus

XII

Submandibular duct

Lingual A.

Hyoid

Hyoglossus

Sublingual gland

Genioglossus

Geniohyoid

Fig. 204   The relationship of the lingual nerve to Wharton's duct.

roots of the 3rd molar tooth. The nerve then passes forwards to the side of the tongue, crossing in turn the lateral aspects of styloglossus, hyoglossus and genioglossus, deep to mylohyoid and above the deep portion of the submandibular salivary gland. In this part of its course, the nerve bears its well-known relationship to Wharton's duct, looping below and round the submandibular duct from the lateral to the medial side as the nerve passes forwards (Fig. 204). The terminal ramifications of the lingual nerve lie directly under the mucosa of the tongue.

The lingual nerve supplies the mucous membrane of the anterior two-thirds of the tongue, and of the side wall and floor of the mouth. In addition, the nerve transmits fibres from the chorda tympani that are secreto-motor to the submandibular and sublingual salivary glands and that convey taste sensation from the anterior two-thirds of the tongue.

The *inferior alveolar nerve* (Fig. 202) is the largest branch of the mandibular nerve. It travels downwards between the medial and lateral pterygoid muscles immediately posterior to the lingual nerve, then, at the lower margin of the lateral pterygoid, it passes in company with the inferior alveolar vessels between the sphenomandibular ligament and the ramus of the mandible to enter the mandibular foramen. The nerve traverses the mandibular canal, supplying branches to each molar and premolar tooth. Between the roots of the 1st and 2nd premolars, the mandibular canal divides into two tunnels – the incisive canal, which continues forwards beneath the canine and incisor teeth, and the mental canal, which runs upwards, laterally and *backwards* to open at the mental foramen. The inferior alveolar nerve similarly bifurcates: the incisive branch passes forwards and supplies the canine and incisor teeth; the mental branch sprays out to supply the chin and the skin and mucous membrane of the lower lip.

Before entering the mandibular foramen, the inferior alveolar nerve gives off its *mylohyoid branch*, which runs downwards and forwards, sandwiched between the medial pterygoid and the inner aspect of the mandible (where it produces the mylohyoid groove), to pierce mylohyoid. It supplies this muscle and, in addition, the anterior belly of digastric.

CLINICAL NOTE

Inferior alveolar nerve block

This nerve is best injected from inside the mouth. In this region, the nerve lies lateral to the sphenomandibular ligament and medial to the ramus of the mandible. The injection is performed at a point immediately medial to the anterior border of the ramus of the mandible approximately 1 cm above the occlusal surface of the 3rd molar tooth. The syringe should be parallel to the body of the mandible and occlusal surfaces. It is easiest to perform from the opposite side with the mouth open. Care should be taken when using dental syringes with cartridges, as aspiration is not possible with these devices.

Anaesthesia is produced in the lower teeth, the skin and mucosa of the lower lip and often, because of spread of the anaesthetic solution, there is loss of sensation of the side of the tongue owing to involvement of the lingual nerve.

If anaesthesia of the lower lip alone is required, this can be obtained by injecting local anaesthetic solution into the mental foramen of the mandible below the 2nd premolar tooth, mid-way between the upper and lower borders of the mandible.

Two ganglia are intimately connected with the mandibular nerve; these are the submandibular and otic ganglia.

Submandibular ganglion

The *submandibular ganglion* is suspended from the lower aspect of the lingual nerve as this passes across the superficial surface of the hyoglossus. It has parasympathetic, sympathetic and sensory connections.

Its *parasympathetic* supply originates in the superior salivary nucleus of VII, passes into the nervus intermedius and joins the main trunk of the facial nerve. These fibres compose the chorda tympani nerve (see page 276), by which they are conveyed to the lingual nerve; they carry the secretomotor supply to the submandibular and sublingual salivary glands.

Sympathetic fibres are transmitted from the superior cervical ganglion via the plexus on the facial artery; they are vasoconstrictor to the submandibular and sublingual glands.

The *sensory component* is contributed by the lingual nerve itself, which provides sensory fibres to these salivary glands and also to the mucous membrane of the floor of the mouth.

Otic ganglion

The *otic ganglion* is unique among the four ganglia associated with the trigeminal nerve in having motor as well as parasympathetic, sympathetic and sensory components. It lies immediately below the foramen ovale as a close medial relationship to the mandibular nerve.

Its *parasympathetic* fibres are derived from the inferior salivary nucleus of the glossopharyngeal nerve. The tympanic branch of this nerve gives off the lesser petrosal nerve, which reaches the ganglion either through the foramen ovale or through the canaliculus innominatus on the medial side of the foramen spinosum. These fibres relay in the ganglion and pass to the auriculotemporal nerve; they are secretomotor to the parotid gland.

The *sympathetic* fibres are derived from the superior cervical ganglion along the plexus that surrounds the middle meningeal artery and sensory fibres arrive from the auriculotemporal nerve; they are vasoconstrictor and sensory to the parotid gland, respectively.

Motor fibres pass to the ganglion from the nerve to the medial pterygoid (a branch of the mandibular nerve). These fibres supply tensor tympani and tensor palati.

Clinically, section of the whole trigeminal nerve results in unilateral anaesthesia of the skin of the face, the anterior part of the scalp and the auricle, and of the mucous membranes of the nose, mouth and anterior two-thirds of the tongue. In addition, there is paralysis and wasting of the muscles of mastication. Lesions of the separate divisions of the nerve give rise to sensory loss in their appropriate areas of supply.

The zones of innervation are summarized in Fig. 205.

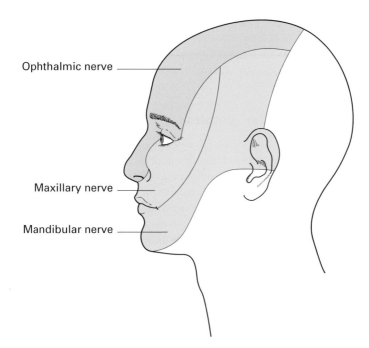

Ophthalmic nerve

Maxillary nerve

Mandibular nerve

Fig. 205 Areas of the face supplied by the three divisions of the trigeminal nerve.

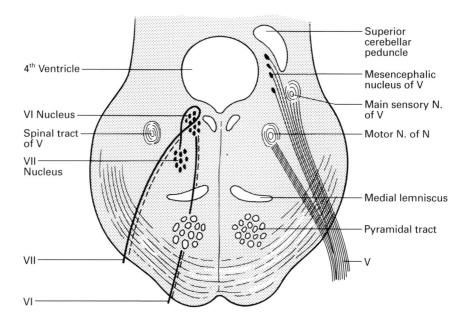

Fig. 206 Transverse section of the pons at the level of the right nucleus of VI to show the intrapontine course of the facial nerve and of the trigeminal nerve on the left.

The abducent nerve (VI)

The abducent nerve, like the trochlear nerve, supplies but one eye muscle: the lateral rectus. Its nucleus is part of the somatic efferent column and lies immediately deep to the floor of the 4th ventricle in the upper part of the pons (Fig. 206); its central connections are considered on page 251. From this nucleus, fibres pass through the pontine tegmentum to emerge on the base of the brain at the junction of the pons and medulla. The nerve then passes forwards to enter the cavernous sinus (Fig. 186). Here, it lies lateral to the internal carotid artery and medial to the IIIrd, IVth and Vth cranial nerves. Passing through the tendinous ring just below the IIIrd nerve, it enters the orbit to pierce the deep surface of the lateral rectus (Fig. 183).

On account of its long and oblique intracranial course, the VIth nerve is frequently involved in injuries to the base of the skull. When damaged, it gives rise to diplopia and a convergent squint. Abduction of the eye (lateral deviation) is impossible on the affected side.

The facial nerve (VII)

The facial nerve supplies the muscles of facial expression, conveys secretomotor fibres to the lacrimal gland, and to the submandibular and sublingual salivary glands, and transmits taste fibres from the anterior two-thirds of the tongue. Separate pontine nuclei are responsible for each of these three functions.

The motor nucleus belongs to the branchial efferent column and is situated within the reticular formation of the lower pons ventromedial to the nucleus of the spinal tract of V; it receives corticobulbar fibres from the motor cortex. The lower part of the nucleus, which controls the upper facial muscles, receives crossed as well as uncrossed cortical fibres; the upper nuclear cells, which control the lower face, receive contralateral fibres only. It is well known that unilateral lesions of the motor cortex, e.g. a stroke, affect only the lower part of the contralateral side of the face.

The secretomotor fibres originate in the superior salivatory nucleus of the general visceral efferent column, which is situated on the reticular formation of the pons dorsilateral to the motor nucleus of VII.

Fibres transmitting taste impulses have their first cell station in the geniculate ganglion, whose central connection is with the upper part of the nucleus of the tractus solitarius (special visceral afferent column). From this nucleus, fibres cross to the opposite lateral nucleus of the thalamus and thence to the face region of the postcentral (sensory) cortex.

From the motor nucleus, efferent fibres of the facial nerve run a devious course over the nucleus of the abducent nerve (Fig. 206), where they form an elevation in the floor of the 4th ventricle known as the facial colliculus, then downwards and forwards to emerge from the lower border of the pons between the olive and the inferior cerebellar peduncle. The sensory fibres emerge as a separate and smaller root, termed the *nervus intermedius*, which, at its origin, lies sandwiched between the motor root of VII medially and the auditory nerve (VIII) laterally (Fig. 179).

The two roots of VII pass together with the auditory nerve into the internal auditory meatus, at the bottom of which they leave VIII and enter the facial canal. Here, they run laterally over the vestibule before reaching the medial wall of the epitympanic recess, then bend sharply backwards over the promontory of the middle ear. This bend, the genu of the facial nerve, marks the site of the facial ganglion and the point at which the secretomotor fibres for the lacrimal gland leave to form the greater petrosal nerve. The facial nerve then passes downwards in the bony posterior wall of the tympanic cavity, to reach the stylomastoid foramen (Fig. 207).

Just before entering this foramen, the facial nerve gives off the *chorda tympani*, which pierces the posterior wall of the tympanic cavity close to the deep surface of the ear drum. It runs forwards over the pars flaccida of the tympanic membrane and the neck of the malleus, lying immediately beneath the mucous membrane throughout its course. It passes out through the front of the middle ear by piercing the bone at a canaliculus at the inner end of the petrotympanic fissure. It emerges from this fissure to join the lingual nerve approximately 2.5 cm below the base of the skull. Through the chorda tympani, taste fibres are conveyed from the anterior two-thirds of the tongue and secretomotor fibres reach the submandibular ganglion.

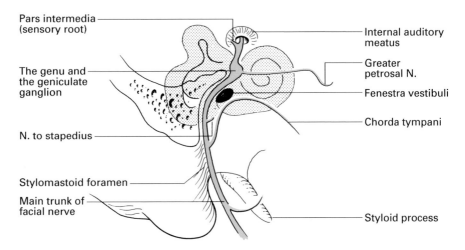

Pars intermedia (sensory root)

Internal auditory meatus

The genu and the geniculate ganglion

Greater petrosal N.

Fenestra vestibuli

Chorda tympani

N. to stapedius

Stylomastoid foramen

Main trunk of facial nerve

Styloid process

Fig. 207 Distribution of the facial nerve within the temporal bone.

Once it emerges from the stylomastoid foramen, the facial nerve is entirely motor in function. The nerve trunk gives off three branches before dividing into its terminal ramifications; these branches are:

1 the *posterior auricular nerve*, which runs backwards over the mastoid process, gives off an auricular branch to the extrinsic muscles of the ear and continues backwards as the occipital branch to the occipital belly of occipitofrontalis;

2 the *digastric branch*, to the posterior belly of digastric;

3 the *stylohyoid branch*, to the stylohyoid muscle.

The trunk of the facial nerve winds lateral to the styloid process, the external carotid artery and the retromandibular vein; here, it can be exposed surgically in the cleft between the mastoid process and the bony part of the external auditory meatus. Just beyond this point, the nerve plunges into the posterior aspect of the parotid gland and bifurcates almost immediately into two divisions, the temporofacial and cervicofacial; occasionally, this bifurcation occurs before the nerve enters the gland.

This unique circumstance of a nerve traversing a gland is explained embryologically. The parotid gland develops as a buccal diverticulum into the crotch formed by the two major divisions of the facial nerve. As the gland enlarges, it overlaps these nerve trunks, the superficial and deep parotid lobes fuse and the nerve comes to lie buried within the gland. The fanciful comparison made of the relationship of the facial nerve and the two lobes of the parotid with the sandwich filling between two slices of bread is not valid, since the two lobes of the parotid become intimately fused around and between the branches of the nerve.

The temporofacial division gives rise to the temporal and zygomatic branches, the smaller cervicofacial division splits into the mandibular and cervical branches while the intermediate buccal branch may originate from either division (Fig. 208). The two divisions may remain completely

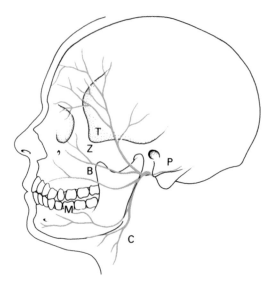

Fig. 208 Distribution of the facial nerve: T, temporal; Z, zygomatic; B, buccal; M, mandibular; C, cervical; P, posterior auricular branch.

separate from each other within the parotid, form a plexus of intermingling connections or, most usually, display a number of cross-connections that can be divided without jeopardy during dissection.

The *temporal branches* cross the arch of the zygoma and supply the muscles of the ear, the frontal belly of occipitofrontalis and orbicularis oculi.

The *zygomatic branches* run across the zygoma to supply orbicularis oculi.

The *buccal branches* pass horizontally forwards to innervate buccinator and the labial musculature.

The *mandibular branch* runs deep to platysma below the angle of the jaw, crosses superficially to the submandibular gland in the digastric triangle, and then runs forwards over the surface of the mandible to supply the muscles of the lower lip and chin. It is particularly liable to injury and may be damaged in the course of excision of the submandibular gland, block dissection of the neck or other operations in this neighbourhood. Since depressor anguli oris is paralysed in consequence, there is a characteristic elevation of the corresponding side of the mouth.

The *cervical branch* passes downwards and forwards into the neck and supplies platysma. In their course through the parotid, the branches of the facial nerve are superficial to the other two structures that traverse the gland, the retromandibular vein (formed by the junction of the superficial temporal and maxillary veins) and the external carotid artery (Fig. 209). The terminal branches of the nerve all emerge from the margins of the parotid; none emerges from the superficial aspect of the gland, which can therefore be completely exposed by the surgeon with impunity.

Damage to the facial nerve or its central pathway results in facial palsy. It is important to distinguish between nuclear and infranuclear facial palsies on the one hand, and supranuclear palsies on the other. Both nuclear and infranuclear palsies result in a facial paralysis that is complete and that affects all the muscles on one side of the face. In supranuclear palsies, there

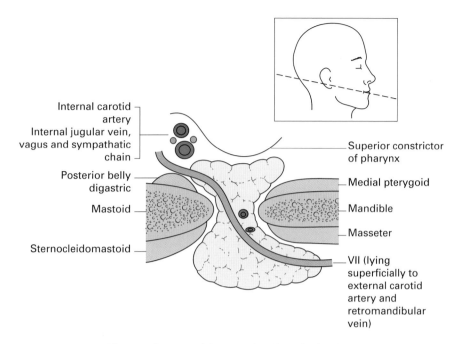

Internal carotid artery
Internal jugular vein, vagus and sympathatic chain
Posterior belly digastric
Mastoid
Sternocleidomastoid

Superior constrictor of pharynx
Medial pterygoid
Mandible
Masseter
VII (lying superficially to external carotid artery and retromandibular vein)

Fig. 209 Horizontal section of the parotid to show the facial nerve traversing the gland. The line of section is shown.

is no involvement of the muscles above the palpebral fissure, since the portion of the facial nucleus supplying these muscles receives fibres from both cerebral hemispheres. Furthermore, in such cases, the patient may involuntarily use the facial muscles but will be unable to do so on request.

Supranuclear facial palsies most frequently result from vascular involvement of the corticobulbar pathways, e.g. in cerebral haemorrhage. Nuclear palsies may occur in poliomyelitis or other forms of bulbar paralysis, while infranuclear palsies may result from a variety of causes, including compression within the cerebellopontine angle (as by an acoustic neuroma), fractures of the temporal bone and invasion by a malignant parotid tumour. However, by far the commonest cause of facial paralysis is Bell's palsy, which is of unknown, probably viral, aetiology.

When the intracranial part of the nerve is affected or when it is involved in fractures of the base of the skull, there is usually an associated loss of taste over the anterior two-thirds of the tongue (chorda tympani involvement) and an associated loss of hearing (VIIIth nerve damage).

The auditory (vestibulocochlear) nerve (VIII)

The auditory nerve consists of two sets of fibres, *cochlear* and *vestibular*, hence its alternative title, the vestibulocochlear nerve. The cochlear fibres

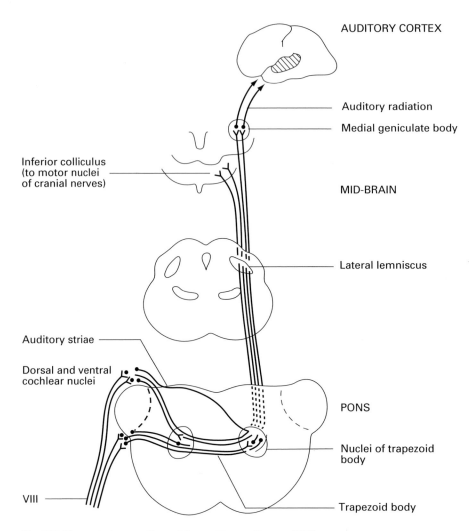

AUDITORY CORTEX

Auditory radiation

Medial geniculate body

Inferior colliculus
(to motor nuclei
of cranial nerves)

MID-BRAIN

Lateral lemniscus

Auditory striae

Dorsal and ventral
cochlear nuclei

PONS

Nuclei of trapezoid
body

VIII

Trapezoid body

Fig. 210 The central connections of the auditory pathway of VIII.

(concerned with hearing) represent the central processes of the bipolar spiral ganglion cells of the cochlea, which traverse the internal auditory meatus to reach the lateral aspect of the medulla, where they terminate in the
dorsal and ventral cochlear nuclei. The majority of the efferent fibres from
these nuclei cross to the opposite side, those from the dorsal nucleus forming the *auditory striae* in the floor of the 4th ventricle, those from the ventral
nucleus forming the *trapezoid body* in the ventral part of the pons (Fig. 210).
Most of these efferent fibres terminate in nuclei associated with the trapezoid body, either on the same or on the opposite side, and then ascend
in the lateral lemniscus either to the inferior colliculus or to the medial
geniculate body; from the former, fibres reach the motor nuclei of the cranial nerves and form the pathway of auditory reflexes, and from the latter,

fibres sweep laterally in the auditory radiation to the auditory cortex on the superior temporal gyrus.

The vestibular fibres (concerned with equilibrium) enter the medulla just medial to the cochlear division and terminate in the vestibular nuclei. Many of the efferent fibres from these nuclei travel to the cerebellum in the inferior cerebellar peduncle, together with fibres that bypass vestibular nuclei. Other vestibular connections are to the nuclei of cranial nerves III, IV, VI and XI via the medial longitudinal bundle, and to the upper cervical cord via the vestibulospinal tract. These connections bring the eye and neck muscles under reflex vestibular control.

Lesions of the cochlear division result in deafness that may or may not be accompanied by tinnitus. Apart from injury to the cochlear nerve itself, unilateral lesions of the auditory pathway do not greatly affect auditory acuity because of the bilaterality of the auditory projections. Temporal lobe tumours may give rise to auditory hallucinations if they encroach upon the auditory radiation or superior temporal gyrus. Lesions of the vestibular division of the labyrinth or of the vestibulocerebellar pathway result in vertigo, ataxia and nystagmus.

The glossopharyngeal nerve (IX)

The glossopharyngeal nerve contains sensory fibres for the pharynx, the tonsillar region and the posterior one-third of the tongue (including the taste buds), motor fibres for the stylopharyngeus muscle and secretomotor fibres for the parotid gland. It also innervates the carotid sinus and body.

Correlated with the varied functions of this nerve, we find that it possesses four nuclei of origin in the brainstem:

1 the rostral part of the nucleus ambiguus is 'borrowed' from the vagus, and constitutes the branchial efferent supply to stylopharyngeus, which is a 3rd branchial arch derivative;

2 the inferior salivatory nucleus, situated immediately rostral to the dorsal motor nucleus of the vagus in the general visceral efferent column, supplies the parasympathetic secretomotor fibres to the parotid gland;

3 taste fibres pass to the nucleus of the tractus solitarius (shared with the chorda tympani branch of VII and the taste fibres of X), representing the special visceral afferent column;

4 fibres of general sensation from the pharynx, tonsil and posterior one-third of the tongue end in the dorsal sensory nucleus of the vagus in the general visceral afferent column.

The glossopharyngeal nerve emerges from the upper part of the medulla by four or five rootlets along the groove between the olive and the inferior cerebellar peduncle (Fig. 179). It first passes forwards and laterally, then leaves the skull by bending sharply downwards through the jugular foramen. Here, it lies in a separate dural sheath in front of the vagus and accessory nerves which, in turn, are anterior to the internal jugular vein.

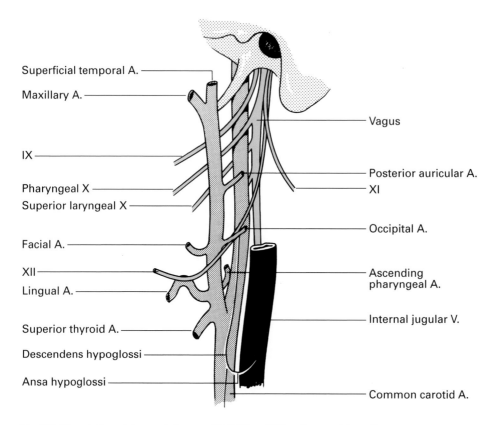

Superficial temporal A.

Maxillary A.

Vagus

IX

Posterior auricular A.

Pharyngeal X

Superior laryngeal X

XI

Occipital A.

Facial A.

XII

Ascending
pharyngeal A.

Lingual A.

Internal jugular V.

Superior thyroid A.

Descendens hypoglossi

Ansa hypoglossi

Common carotid A.

Fig. 211 The relation of the cranial nerves IX, X, XI and XII to the carotid arteries and the jugular vein.

Within the jugular foramen, the glossopharyngeal nerve bears a superior and an inferior ganglion, which are the first cell stations for the fibres of taste and common sensation. Also within the foramen, the nerve gives off its tympanic branch.

Below the jugular foramen the nerve courses downwards and forwards between the internal carotid artery and the internal jugular vein deep to the styloid process and its muscles. It then curves forwards between the internal and external carotid arteries (Fig. 211), across the stylopharyngeus muscle, to enter the pharynx between the superior and middle constrictors. Here, it breaks up into its terminal branches that supply the posterior one-third of the tongue (both common sensation and taste), the mucous membrane of the pharynx and the tonsil.

The *tympanic branch* supplies the tympanic cavity and is continued as the lesser petrosal nerve, which conveys the preganglionic parasympathetic fibres to the otic ganglion (parotid secretomotor fibres; see page 274).

The *carotid branch* arises just below the skull and runs down on the internal carotid artery to supply the carotid body and sinus. This small twig serves as the afferent limb of the pressor-receptor and chemoreceptor reflexes from the carotid sinus and body, respectively.

The *carotid sinus* is a bulge at the start of the internal carotid artery. Here, the arterial wall is thin and has a particularly rich nerve supply from the glossopharyngeal nerve. It is the principal pressor-receptor in the body and its stimulation causes a reflex decrease in blood pressure and slowing of the heart.

The *carotid body* is a small, oval, reddish-brown structure approximately 5 mm in length that lies deep to the bifurcation of the common carotid artery. Together with the aortic body, it is sensitive to chemical changes in the blood, particularly to variations of CO_2 and O_2 tension and to increased H^+ concentration. Histologically, the gland is made up of polyhedral cells, among which are scattered chromaffin cells. There is a rich sinusoidal blood supply together with dense ramifications of nerve fibres from the glossopharyngeal nerve. The blood flow per gram of tissue is higher than that of any other organ and gives it a unique opportunity to monitor changes in blood H^+ and O_2 content.

Complete section of the glossopharyngeal nerve results in sensory loss in the pharynx, loss of taste and common sensation over the posterior one-third of the tongue, some pharyngeal weakness and loss of salivation from the parotid gland. However, such lesions are obviously difficult to detect and rarely occur as an isolated phenomenon since there is so often associated involvement of the vagus nerve or its nuclei.

Glossopharyngeal neuralgia is an extremely distressing condition in which severe pain in the tonsillar area is triggered by yawning or, in advanced cases, by mastication. It is amenable to *glossopharyngeal nerve block*.

CLINICAL NOTE

Glossopharyngeal nerve block

This block is conducted for the treatment of glossopharyngeal neuralgia and in the palliative care of patients with head and neck malignancy. It is performed by infiltrating the nerve as it emerges from the jugular foramen before it turns deep to the styloid process. A line is drawn from the mastoid process to the angle of the mandible. At the mid-point of this line, a needle is advanced perpendicular to the skin. The styloid process should be found within 3 cm. The needle is walked posteriorly and injection should be made as soon as bony contact is lost. As the anaesthetic almost invariably spreads, other cranial nerves may be affected, including X, XI and XII. However, apart from occasional weakness of the sternomastoid and trapezius, the effects of this block cause remarkably little disturbance to a patient. Tachycardia and dysphonia may indicate vagal block. Under no circumstances should bilateral block be attempted.

The vagus nerve (X)

The vagus nerve is the largest and most widely distributed of the cranial nerves; it is also unique among them in being asymmetrical. It conveys motor, sensory and secretomotor fibres and its distribution can be summarized thus:

1 *Motor* to:
   a larynx;
   b bronchial muscles;
   c alimentary tract (as far as the splenic flexure);
   d heart (cardio-inhibitory).
2 *Sensory* to:
   a dura;
   b external auditory meatus;
   c respiratory tract;
   d alimentary tract (as far as the ascending colon);
   e heart;
   f epiglottis (gustatory).
3 *Secretomotor* to:
   a bronchial mucus glands;
   b alimentary tract and its adnexae.

The vagus has its origin in three medullary nuclei: the dorsal nucleus, the nucleus ambiguus and the nucleus of the tractus solitarius (Fig. 212).

The *dorsal nucleus of the vagus*, situated below the floor of the 4th ventricle in the central grey matter of the caudal part of the medulla, is a general

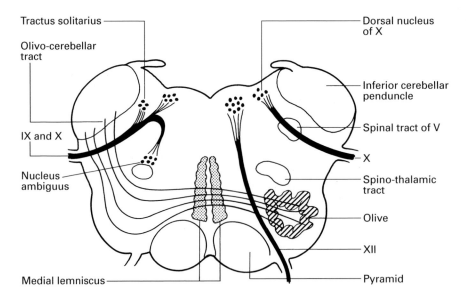

Fig. 212 Transverse section through the medulla at the level of the caudal part of the 4th ventricle to show the nuclei of X and XII.

visceral, mixed sensory and motor centre. It is the cell station for motor fibres to the heart, bronchi and alimentary tract and also receives sensory fibres from the larynx and lungs, the pharynx and alimentary tract, and from the heart.

The *nucleus ambiguus* of the branchial efferent column provides the motor fibres of IX, X and XI, which supply the voluntary muscles of branchial origin of the pharynx, larynx and palate; it lies deeply within the reticular formation of the medulla.

The *nucleus of the tractus solitarius* (the special afferent column) is concerned with afferent gustatory impulses and is situated in the central grey matter of the medulla. Its anterior part receives fibres from the chorda tympani branch of VII, its middle part from IX and its posterior part from X. This last receives taste fibres that arrive from the epiglottis and valleculae along the internal branch of the superior laryngeal nerve.

From the medulla, the vagus emerges as about 10 rootlets in series with the glossopharyngeal nerve in the posterolateral sulcus between the olive and the inferior cerebellar peduncle (Fig. 179). The nerve rapidly unites into a single trunk that passes through the jugular foramen in a common dural sheath with the accessory nerve (XI), separated from IX by a fibrous septum.

(Note that, from before backwards, the jugular foramen transmits the inferior petrosal sinus; IX, X and XI, one behind the other; and lastly the internal jugular vein.)

The vagus bears a pair of ganglia, one within the foramen and the other one emerging from it; these contain the unipolar cells of the sensory component of the nerve, just as the ganglia on the posterior nerve roots bear the relay cells of the spinal sensory nerves. The superior (or jugular) ganglion communicates with filaments from IX and the superior cervical sympathetic ganglion. The inferior ganglion (or nodosum) communicates with XII and with the loop that connects the 1st and 2nd cervical nerves.

Course and relations

The vagus descends through the neck within the carotid sheath (Fig. 37), lying between and slightly posterior to the internal jugular vein and internal carotid artery above and then more distally between the internal jugular vein and the common carotid artery (Fig. 211). The cervical sympathetic chain lies behind the carotid sheath, which is separated by the prevertebral fascia from longus capitis and longus cervicis. From the root of the neck downwards, the relations of the right and left vagus are quite dissimilar.

The *right vagus* crosses the first part of the subclavian artery and gives off the recurrent laryngeal nerve at the lower border of this vessel (Fig. 213), then, passing behind the right brachiocephalic vein, the nerve descends into the thorax against the lateral aspect of the trachea, on which it is crossed by the vena azygos, the only structure to separate the nerve from the lung and pleura (Fig. 54). The vagus then passes behind the root of the lung, where it breaks up to form (together with sympathetic fibres) the right posterior pulmonary plexus, whose branches enter the lung at its

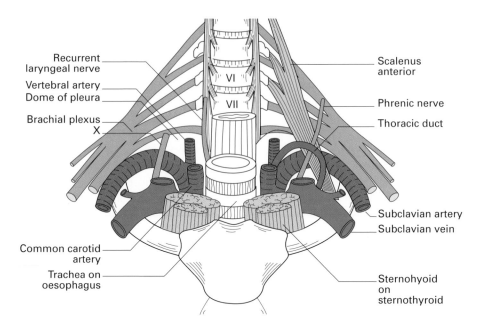

Fig. 213 The root of the neck to show the relationships of X and the right recurrent laryngeal nerve to the subclavian artery.

hilum. From this plexus, two or more cords emerge onto the posterior aspect of the oesophagus, receive a contribution from the left vagus and form a posterior oesophageal plexus, from which the right vagus trunk (now containing fibres from both sides) is reconstituted.

The right (or posterior) vagus emerges into the abdomen through the diaphragmatic oesophageal hiatus (Fig. 62), where the nerve lies to the right, well behind the oesophagus. It gives branches to both the anterior and the posterior aspects of the upper part of the body of the stomach, but the bulk of the nerve forms the coeliac branch; this runs along the left gastric artery to the coeliac ganglia and is distributed to the intestine and its associated organs, to the kidneys and to the suprarenal glands (Fig. 214b).

The *left vagus* enters the thorax between the left carotid and left subclavian arteries, behind the left brachiocephalic vein. It crosses the aortic arch more lateral and more posterior to the point at which the phrenic nerve passes over the arch (Fig. 55), the two nerves being separated by the left superior intercostal vein. At the lower border of the arch, the vagus gives off its recurrent laryngeal branch. Passing now behind the lung root, the vagus breaks up into the left posterior pulmonary plexus, from which two or more cords descend onto the front of the oesophagus; these receive filaments from the right vagus to form the anterior oesophageal plexus. From this, the anterior vagus emerges as a single trunk, which contains fibres derived from both sides and which passes through the oesophageal hiatus

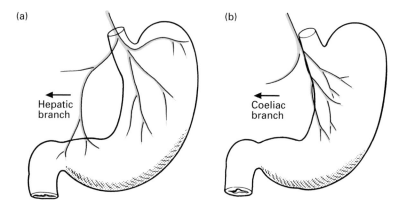

Fig. 214 The vagal supply to the stomach: (a) anterior vagus; (b) posterior vagus.

in front of, and (in contrast to the posterior vagus) in very close apposition to, the oesophagus.

The anterior vagus supplies branches to the cardia and to the lesser curvature of the stomach and also gives off a large hepatic branch (Fig. 214a), which, in turn, donates a pyloric branch to the upper border of the pylorus and to the antral region of the stomach termed the *nerve of Latarjet*.

There are two reasons for the asymmetry between the right and left vagus nerves. The development of the aortic arch and its branches accounts for the differences in relationships at the root of the neck and for the courses taken by the recurrent laryngeal nerves (Fig. 79). The embryological rotation of the stomach, in which the left lateral aspect swings round to become the definitive anterior wall during the formation of the lesser sac, is responsible for the left vagus coming to lie anteriorly at the oesophageal hiatus, and the right vagus correspondingly taking up a posterior position in relation to the stomach.

The branches and distribution of the vagus nerve

The widespread distribution of the vagal branches is best classified regionally according to their sources of origin:

1 Jugular fossa:
   a meningeal branch;
   b auricular branch.
2 Neck:
   a pharyngeal branch;
   b superior laryngeal nerve;
   c right recurrent laryngeal nerve;
   d cardiac branches.
3 Thorax:
   a cardiac branches;
   b left recurrent laryngeal nerve;
   c anterior and posterior pulmonary branches;

    d pericardial branches;

    e oesophageal branches.

4 Abdomen:

    a gastric branches;

    b hepatic branch;

    c coeliac branches.

The *meningeal branch* arises from the superior ganglion, passes backwards through the jugular foramen and supplies the dura.

The *auricular branch* also originates from the superior ganglion, then enters a tiny canal on the lateral wall of the jugular fossa that leads it into the temporal bone, from which the nerve emerges again between the mastoid process and the tympanic plate. It supplies the medial aspect of the auricle, the external auditory meatus and the outer surface of the tympanic membrane. It communicates with the facial nerve both in the petrous temporal bone and again with the posterior auricular branch of VII on emerging from the bone.

This is the 'alderman's nerve' – cold water or other stimuli applied to the external ear are said to stimulate the appetite; the anatomical basis for this is the vagal innervation of both the external ear and the gastric secretory and emptying mechanism. Irritation of the ear drum may indeed produce vomiting, perhaps as a vagal reflex, and certainly acute otitis in children may present as abdominal pain.

The *pharyngeal branch* (Fig. 211) arises from the inferior ganglion of the vagus. The standard textbook description is that its fibres chiefly derive from the cranial root of the accessory (XI) nerve. A recent study has shown that these fibres are, indeed, truly vagal, derived from the lower part of the nucleus ambiguus and leaving the medulla in four or five rootlets (see accessory nerve; page 289). The nerve passes downwards and forwards between the internal and external carotid arteries to reach the middle constrictor of the pharynx, where it helps in the formation of the *pharyngeal plexus*. This plexus receives, in addition, the pharyngeal branch of IX and sympathetic fibres from the superior cervical ganglion.

Branches from the pharyngeal plexus supply:

1 the superior, middle and inferior constrictors of the pharynx;
2 palatoglossus, palatopharyngeus and levator palati, i.e. all the palatal muscles apart from the tensor palati, whose nerve supply is derived from the mandibular branch of V;
3 sensory fibres to the pharyngeal mucosa.

The *superior* and *recurrent laryngeal nerves* are considered in the section on the larynx (see page 37).

The *cardiac branches* arise in the neck and mediastinum. The cervical branches are usually two in number: one from the upper vagus, the other originating in the root of the neck. On the right side, these descend behind the subclavian artery, along the trachea to join the deep cardiac plexus (see page 232). On the left, these branches accompany the vagus into the thorax; the upper then tracks along the trachea to the deep cardiac plexus; the lower crosses in front of the aortic arch to reach the superficial plexus.

Other cardiac branches arise from the vagus within the mediastinum and also from both the recurrent laryngeal nerves; these all pass to the deep cardiac plexus.

The *pulmonary branches* are divided into an anterior and a posterior group. The anterior branches are two or three in number, originate just above the lung hilum and enter the anterior pulmonary plexus. The posterior nerves are larger and more numerous; they help form the posterior pulmonary plexus. Both plexuses also have sympathetic components, and are considered on page 240.

The distributions of the oesophageal, gastric, coeliac and hepatic branches have already been described above.

The communications of the vagus

A number of communications between the vagus and other nerves have been given in the above description. They can be summarized thus:

1 The trunk and its ganglia with IX, XI, XII, the superior cervical sympathetic ganglion and with the 1st and 2nd cervical nerves.
2 The auricular branch with VII.
3 The pharyngeal plexus with IX and the superior cervical ganglion.
4 The cardiac, pulmonary, oesophageal and gastric branches with the sympathetic outflow to the viscera.

The accessory nerve (XI)

The accessory nerve (Figs 127, 179) is conventionally described as having a cranial and a spinal root. According to standard descriptions, the cranial root is formed by a series of rootlets that emerge from the medulla between the olive and the inferior cerebellar peduncle. These rootlets are considered to join the spinal root, travel with it briefly, then separate within the jugular foramen and are distributed with the vagus nerve to supply the musculature of the palate, pharynx and larynx.

A recent, detailed dissection study has demonstrated that all the medullary rootlets that do not join to form the glossopharyngeal nerve (IX) join the vagus nerve at the jugular foramen. All the rootlets that form the accessory nerve arise caudal to the olive and no connections can be demonstrated between the accessory nerve and the vagus in the jugular foramen. The accessory nerve thus has *no* cranial component and consists only of the structure hitherto referred to as the spinal root of the accessory nerve.

This spinal root is formed by the union of fibres from an elongated nucleus in the anterior horn of the upper five cervical segments, which leave the cord midway between the anterior and posterior roots, join, then pass upwards through the foramen magnum. The accessory nerve and the converging rootlets of the vagus nerve then enter the jugular foramen in a shared sheath of dura. The glossopharyngeal nerve enters the jugular foramen anterior to the vagus through a separate dural sheath.

It might be argued that the accessory nerve should no longer be regarded as a 'cranial' nerve. This is a question of semantics; the fact that it does exit through a foramen in the skull can lead us to regard it as a cranial nerve, notwithstanding the fact now established that it has an exclusive origin from the upper cervical spinal cord.

The spinal root passes backwards over (or less often deep to) the internal jugular vein (Fig. 211), crosses the transverse process of the atlas and is itself crossed by the occipital artery to reach the sternocleidomastoid muscle, which it pierces and supplies. It then crosses the posterior triangle of the neck to enter the deep surface of the trapezius approximately 5 cm above the clavicle.

Isolated lesions of the cranial root of the accessory nerve are infrequent; more commonly, it is involved concomitantly with the vagus when it gives rise to paresis of the laryngeal and pharyngeal muscles, resulting in dysphonia and dysphagia.

Division of the fibres of the spinal root (or lesions affecting their cells of origin) results in paresis of the sternocleidomastoid and trapezius muscles. This follows, for example, most block dissections of the lymph nodes of the neck, the nerve being sacrificed in clearing the posterior triangle.

It is easy to draw the surface markings of the accessory nerve; one merely constructs a line from the tragus to a point along the anterior border of the trapezius 5 cm above the clavicle. This line will cross the transverse process of the atlas and also the junction of the upper and middle thirds of the posterior border of sternocleidomastoid.

The hypoglossal nerve (XII)

The hypoglossal nerve supplies all the intrinsic and extrinsic muscles of the tongue (with the exception of palatoglossus). From its nucleus, which is part of the somatic efferent column and which lies in the floor of the 4th ventricle, a series of about a dozen rootlets leaves the side of the medulla in the groove between the pyramid and the olive (Fig. 212). These rootlets unite to leave the skull by way of the anterior condylar, or hypoglossal, canal. Lying at first deep to the internal carotid artery and the jugular vein, the nerve passes downwards between these two vessels to just above the level of the angle of the mandible. Here, it passes forwards over the internal and external carotid arteries (Fig. 211), across the loop of the lingual artery, then inclines upwards and forwards on the hyoglossus, passing deep to the tendon of digastric, the stylohyoid and the mylohyoid. On hyoglossus, the nerve is related to the deep aspect of the submandibular gland and then lies inferior to the submandibular duct and the lingual nerve. It then passes on to genioglossus and ends by being distributed to the muscles of the tongue (Fig. 215).

The hypoglossal nerve receives an important contribution from the anterior primary ramus of the 1st cervical nerve at the level of the atlas. The majority of these C1 fibres pass into the *descendens hypoglossi*, which is

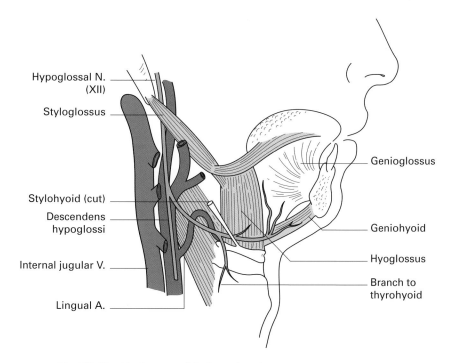

Hypoglossal N.
(XII)

Styloglossus

Stylohyoid (cut)

Descendens
hypoglossi

Internal jugular V.

Lingual A.

Genioglossus

Geniohyoid

Hyoglossus

Branch to
thyrohyoid

Fig. 215 The distal course of the hypoglossal nerve.

given off as the main trunk of the nerve crosses the internal carotid artery. This descending branch passes downwards on the carotid sheath and is joined by *descendens cervicalis*, derived from C2 and C3, to form a loop, the *ansa hypoglossi* (Fig. 126). From this loop, branches pass to supply omohyoid, sternothyroid and sternohyoid. Other C1 fibres are transmitted by the hypoglossal nerve to thyrohyoid and geniohyoid.

Division of the hypoglossal nerve (which may occur, for example, inadvertently when performing a carotid endarterectomy) or lesions involving its nucleus result in ipsilateral paralysis and wasting of the muscles of the tongue. This is detected clinically by deviation of the protruded tongue to the affected side. Supranuclear paralysis (due to involvement of the corticobulbar pathways) leads to paresis, but not atrophy, of the muscles of the contralateral side.

Part 7
Miscellaneous Zones of Interest

The thoracic inlet

The thoracic inlet, viewed either on the skeleton or in the dissecting room, is a surprisingly small space into which are packed the apices of the lungs, the trachea, the oesophagus and the great vascular trunks: the brachiocephalic artery, the brachiocephalic veins, the left common carotid and the left subclavian artery – together with the vagi, the cervical sympathetic chains, the phrenic nerves and the thoracic duct. As a backcloth to the inlet on either side lies the brachial plexus, sandwiched between the scalenus anterior and medius. It is the brachial plexus, of course, that renders this area of intense practical importance to the anaesthetist but, in addition, the anaesthetist may be called upon to block the stellate ganglion or cannulate the internal or external jugular vein (Fig. 213).

Outlines and boundaries

The thoracic inlet is kidney-shaped because of the forward projection into it of the body of the 1st thoracic vertebra; it measures some 10 cm in transverse diameter and 5 cm anteroposteriorly. Its boundaries are the body of the 1st thoracic vertebra, the 1st ribs, their costal cartilages and the upper border of the manubrium sterni.

The inlet slopes downwards sharply from behind forwards, forming an angle of approximately 60° with the horizontal. There is, in fact, a 4 cm difference in height between the anterior and posterior extremities of the inlet, the upper border of the manubrium lying opposite the disc between the 2nd and 3rd thoracic vertebrae (Fig. 49). During quiet respiration, this level hardly varies, but in forced inspiration and expiration the upper border of the manubrium moves about the length of a vertebral body in each direction.

The 1st rib (Figs 216, 217)

The *1st rib* is the key to the important neurovascular relationships of this region. It is unique in being the shortest, flattest and most curvaceous of the ribs. Its extreme flattening and curvature give it broad upper and lower surfaces and sharp outer and inner margins, the latter bearing the scalene tubercle, the site of insertion of the anterior scalene muscle.

The 1st rib has a rounded *head*, with a single *facet* for the body of the 1st thoracic vertebra, a long *neck* and a prominent *tubercle*, which articulates with the transverse process of the 1st thoracic vertebra. Crossing the neck are: medially, the sympathetic trunk; laterally, the large branch of the anterior primary ramus of the 1st thoracic nerve passing to the brachial plexus; between them, the superior intercostal artery (derived from the costocervical trunk).

The *scalene tubercle* provides the insertion for the tendon of scalenus anterior. Immediately in front of this tubercle, the upper surface of the rib bears

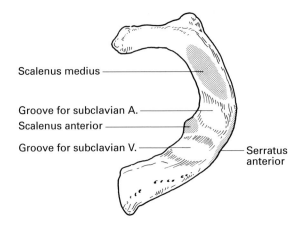

Scalenus medius

Groove for subclavian A.
Scalenus anterior

Groove for subclavian V.

Serratus
anterior

Fig. 216 The 1st rib
viewed from above.

a groove for the subclavian vein; because of the obliquity of the thoracic
inlet, this vessel lies well below (and safely behind) the clavicle.

Behind the scalene tubercle lies a second groove that is for the subclavian
artery and the lower trunk (C8, T1) of the brachial plexus. This groove
is particularly well marked when the subject has a 'post-fixed' brachial
plexus with a large contribution from T2. Immediately behind this groove
is the area of insertion of scalenus medius.

To the inner margin of the 1st rib is attached the suprapleural membrane,
better known as Sibson's fascia. This is a tough sheet of fibrous tissue that
spreads out like a tent from its origin, the transverse process of C7, to form
a protective covering over the cervical pleura. Note that the apex of the
pleura extends approximately 4 cm above the medial third of the clavicle
(Fig. 41). It may be inadvertently punctured during the performance of a
brachial plexus block or subclavian vein cannulation.

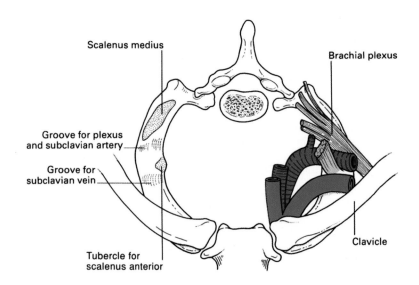

Scalenus medius

Brachial plexus

Groove for plexus
and subclavian artery

Groove for
subclavian vein

Clavicle

Tubercle for
scalenus anterior

Fig. 217 Superior view
of the thoracic inlet and
the structures that pass
across the 1st rib.

Of less practical importance, but to complete this list of 1st rib relations and attachments, the small subclavius muscle arises from the anterior extremity of the upper surface of the rib to become inserted into the under aspect of the clavicle. Serratus anterior and the intercostal muscles of the first space find attachment to the lateral margin of the rib, and its inferior aspect lies against the cervical pleura.

Cervical ribs

Approximately 0.5% of individuals have a cervical rib, which may comprise merely an enlarged costal process of C7, continuing as a fibrous strand to the 1st rib just beyond the scalene tubercle (the commonest manifestation), or a true rib with articular facets that articulate with the body and transverse process of C7, again with a fibrous connection with the 1st rib, or, finally, a complete rib that articulates or fuses with the front of the 1st rib and that has the scalene muscles attached to it.

A cervical rib is usually symptomless, but it may sometimes be associated with either neurological or vascular disturbances. It may seem rather strange, but pressure effects on the brachial plexus are rare in the presence of a completely developed cervical rib. This is because, in such cases, the plexus is usually 'pre-fixed' (its main components being derived from C4–8), and lies clear of the extra rib. Conversely, the plexus may be 'postfixed', from C6 to T2, and this may be associated with an anomalous 1st thoracic rib, which is rudimentary and replaced by a fibrous strand.

The lower cord of the normally constituted brachial plexus may be snared over the fibrous prolongation of an incomplete cervical rib with resultant paraesthesia over the distribution of C8 and T1 (the ulnar border of the forearm and hand), together with weakness and wasting of the small muscles of the hand, especially those of the thenar eminence.

The subclavian artery must, of necessity, arch over a complete cervical rib; it is then unusually prominent in the supraclavicular fossa and is often mistaken for an aneurysm. The artery may be narrowed at this site, and develop a poststenotic dilatation in which thrombosis may occur. Emboli arising from this source may result in the cold, cyanosed upper limb, with absent pulses, claudication pain and perhaps digital gangrene that is sometimes seen in association with a cervical rib.

Surface markings

The 1st rib is a structure of importance in supraclavicular approaches to the brachial plexus such as the subclavian perivascular and Kulenkampff supraclavicular techniques. In the supraclavicular fossa, the trunks of the plexus are closely clumped together between the rib and the overlying skin and can be rolled under the finger at this site in the thin subject. The nerves of the brachial plexus occupy their smallest volume at this level and relatively small volumes of local anaesthetic can produce whole-arm anaesthesia. It is unfortunate that the 1st rib has close relations to other structures that the anaesthetist would not wish to have access to when armed with a syringe of local anaesthetic: the subclavian artery and vein, and the pleura.

Unfortunately, the 1st rib is completely impalpable in many people, and we must rely on other landmarks to find the zone where the plexus crosses the rib. There are two useful guides that lead to the same place immediately above the clavicle where the anaesthetist's needle might be targeted. These are the mid-point of the clavicle, and the point immediately lateral to a finger pressing on the pulsations of the subclavian artery. This pulse is felt where the artery, having emerged from between the scalene muscles, crosses the 1st rib immediately in front of the plexus (Figs 135, 217). Further details on the anatomy of brachial plexus block are given on page 171.

The antecubital fossa

To the anatomist, the antecubital fossa is the space through which the vascular and nervous trunks course into the forearm; to the surgeon, it is the region in which the brachial artery is put into jeopardy by injuries around the elbow; to the anaesthetist, it is the place where he/she is tempted to seek a superficial vein, knowing only too well the dangers of inadvertent intra-arterial or intraneural injection. It is also a convenient area for arterial cannulation and blocks of the four nerves to the lower arm.

Boundaries

The antecubital fossa is a triangle delimited by pronator teres inferomedially, brachioradialis inferolaterally and a line joining the medial and lateral epicondyles of the humerus above.

Roof (Fig. 218)

The fossa is roofed by deep fascia reinforced by the bicipital aponeurosis. On this deep fascia lies the median cubital vein crossed superficially (or sometimes deeply) by the medial cutaneous nerve of the forearm, which is here occasionally damaged at venepuncture. Laterally lie the cephalic vein and the lateral cutaneous nerve of the forearm, while medially courses the basilic vein.

Contents (Figs 219, 220)

If the muscular walls of the fossa are retracted, the following structures can be identified in turn from the medial to lateral side.
1 The median nerve.
2 The brachial artery, which bifurcates at the level of the neck of the radius into its terminal radial and ulnar branches.
3 The biceps tendon.
4 The radial nerve, giving off its posterior interosseous branch.

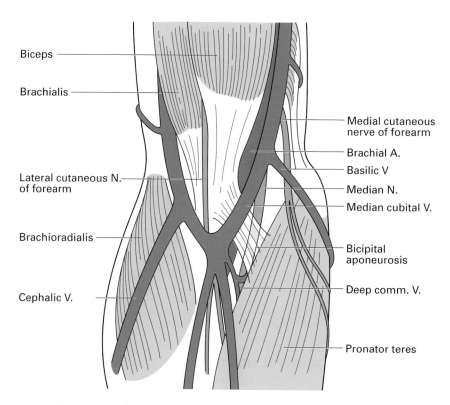

Biceps

Brachialis

Medial cutaneous nerve of forearm

Brachial A.

Basilic V

Lateral cutaneous N. of forearm

Median N.

Median cubital V.

Brachioradialis

Bicipital aponeurosis

Cephalic V.

Deep comm. V.

Pronator teres

Fig. 218 Superficial dissection of the antecubital fossa.

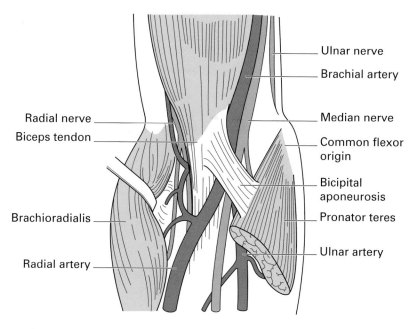

Ulnar nerve

Brachial artery

Radial nerve

Biceps tendon

Median nerve

Common flexor origin

Bicipital aponeurosis

Brachioradialis

Pronator teres

Ulnar artery

Radial artery

Fig. 219 Deep dissection of the antecubital fossa.

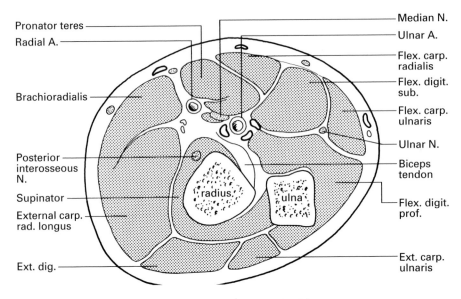

Pronator teres

Radial A.

Brachioradialis

Posterior
interosseous
N.

Supinator

External carp.
rad. longus

Ext. dig.

Median N.

Ulnar A.

Flex. carp.
radialis

Flex. digit.
sub.

Flex. carp.
ulnaris

Ulnar N.

Biceps
tendon

Flex. digit.
prof.

Ext. carp.
ulnaris

radius

ulna

Fig. 220 Transverse section through the apex of the antecubital fossa.

Structures of clinical importance

Superficial veins (Figs 219, 221)

The *cephalic vein* drains tributaries from the radial border of the forearm,
ascends over the lateral side of the antecubital fossa and then lies in a
groove along the lateral border of the biceps. The vein dives beneath the
deep fascia at the lower border of pectoralis major, lies in a groove between
this muscle and deltoid, then finally pierces the clavipectoral fascia to enter
the axillary vein. This groove between pectoralis major and deltoid is a
conveniently identifiable site for a cut-down when a superior vena caval
infusion is required.

CLINICAL NOTE

Catheters inserted into the cephalic vein frequently fail to enter the axillary
vein and superior vena cava because of this sharp curve as it passes through
the clavipectoral fascia and because of the valve commonly guarding this
venous junction. Superior vena cava cannulation by this route is said to be
more frequently successful if the right cephalic vein is selected.

The *basilic vein* drains the ulnar side of the forearm and then ascends
along the medial border of the biceps to pierce the deep fascia at the middle
of the upper arm. From here, the vein runs upwards to the lower border of
the axilla, where it is joined by the venae comitantes of the brachial artery
to form the axillary vein.

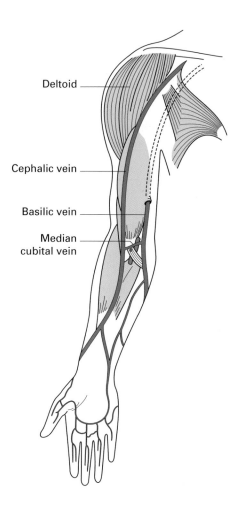

Fig. 221 Superficial veins of the upper limb.

The *median cubital vein* (alternatively termed the median basilic or median cephalic vein) usually arises from the cephalic vein approximately 2.5 cm distal to the lateral epicondyle then runs upwards and medially to join the basilic vein 2.5 cm above the transverse crease of the elbow, giving the rather drunken 'H'-shaped arrangement shown in Fig. 222.

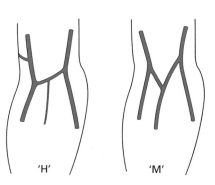

Fig. 222 Variations in the formation of the median cubital vein.

The median cubital vein receives a number of tributaries from the front of the forearm, as well as giving off a deep median vein that pierces the fascial roof of the antecubital fossa to join the venae comitantes of the brachial artery.

A frequent variation in this venous arrangement is for a median forearm vein to bifurcate just distal to the antecubital fossa; one limb then passes to the cephalic, the other to the basilic vein, to give an 'M'-shaped pattern (Fig. 222).

The bicipital aponeurosis (Figs 218, 219)

Arising from the medial border of the lower end of the biceps muscle and its tendon is the flat bicipital aponeurosis. This passes downwards and medially to blend with the deep fascia covering the origin of the forearm flexor muscles. The upper edge of this aponeurosis is somewhat thickened and can be felt quite easily when the forearm is flexed and supinated. The aponeurosis lies as a shield between the brachial artery and the median cubital vein. This fortuitously placed fascial barrier was much appreciated by the barber-surgeons of old, who used the antecubital veins for blood-letting; they named it the *grace à Dieu* fascia (the 'praise be to God' fascia).

The brachial artery (Fig. 219)

The variations in the anatomy of the brachial artery are of great importance to the anaesthetist. They include:

1 The artery may bifurcate high in the arm, even at axillary level, into a main trunk (which continues into the forearm as the common interosseous artery) and a common stem, termed the superficial brachial artery, which divides at a variable level into its radial and ulnar branches (1% of cases).

2 A superficial radial artery may be given off in the upper arm; an anomaly of little practical importance because this vessel continues its course in a manner identical to a normal radial artery (14% of cases).

3 A superficial ulnar artery is given off in the upper arm in 2% of cases and descends nearly always superficially to the common origin of the fore-arm flexors. It then sweeps downwards to the lateral border of flexor carpi ulnaris to take up, in the distal forearm, the relationships of a normal ulnar artery. This superficial ulnar artery may lie beneath the deep fascia throughout its course or may lie subcutaneously, either at the elbow or in the upper forearm.

It is the anomalous ulnar artery in this superficial and subcutaneous position, immediately beneath the median cubital vein and without the protection of the bicipital aponeurosis, that is at greatest risk of accidental puncture when the antecubital vein is selected for an intravenous injection. Fortunately, the superficial ulnar artery (unlike the normal ulnar artery) usually does not give rise to the common interosseous artery; in such cases, the latter is derived from the radial artery. An inadvertent intra-arterial injection into the superficial ulnar artery will therefore usually spare this common interosseous branch and tend to preserve undamaged the main source of arterial blood supply to the forearm muscles.

CLINICAL NOTE

Arterial cannulation sites

Arteries are cannulated for continuous blood pressure monitoring, repeated arterial blood sampling and in situations in which non-invasive blood pressure measurement is not feasible. Although any artery capable of accommodating a small cannula will suffice, some general principles apply: the artery should be away from the site of surgery or disease processes and its flow should not be compromised by the nature of the surgery or disease; the artery should not directly supply the brain because of the chance of embolic damage; the artery should ideally not be an end-artery and the tissue supplied by the artery should have adequate demonstrable collateral flow. For these reasons, arteries at the wrist and ankle are the most popular sites for arterial cannulation: the radial artery at the wrist and the dorsalis pedis artery on the dorsum of the foot. The ulnar artery and the posterior tibial artery are also occasionally used. Cannulation of large end-arteries such as the brachial, axillary and femoral arteries is acceptable if more distal alternatives are not available.

The orbit and its contents

The orbit and its contents are of particular importance to the anaesthetist from the point of view of regional blocks for eye surgery, and in the management of 'blow-out' maxillary fractures and frontal head injuries.

Note: for details of the optic nerve and the visual pathway, see pages 249–251; details of the cranial nerves within the orbit will be found on pages 251–275 (oculomotor nerve, III, pages 251–253; trochlear nerve, IV, pages 253–254; ophthalmic nerve, V', pages 258–261; and abducent nerve, VI, page 275).

The bony orbit

The bony margin of the orbit is quadrilateral in shape, made up of the frontal bone, zygomatic bone and maxilla (Fig. 223). The junction between the frontal process of the zygoma and the zygomatic process of the frontal bone halfway along the lateral margin is palpable as a distinct notch. The supra-orbital notch can be felt in the middle of the superior margin; pressure with the fingernail here is painful because of the supra-orbital nerve (V') lying within the notch. The exception to this is when the notch is replaced by a foramen.

The bony orbital cavity is pyramidal. The medial walls on each side are parallel, separated by the nasal cavity, and approximately 2.5 cm apart. The

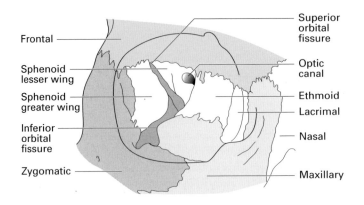

Fig. 223 The bony orbit.

lateral walls of the two cavities are at right angles to each other (Fig. 224). The apex of the pyramid is the optic foramen and its base is the orbital margin. The eyeball is 2.5 cm in diameter, half the length of the orbital cavity, which has a volume of approximately 30 ml. The medial and lateral wall are each about 5 cm in length.

Seven bones go into the complex structure of the bony orbit: the frontal, zygomatic, maxilla, ethmoid, sphenoid, lacrimal and palatine (Fig. 223).

The optic foramen lies between the two roots of the lesser wing and the body of the sphenoid. Lateral to this foramen is a prominent 'V'-shaped fissure. The upper limb of the 'V' is the superior orbital fissure, which separates the lesser from the greater wing of the sphenoid. The lower limb of the 'V', the inferior orbital fissure, separates the greater wing from the maxilla. This fissure continues as the infra-orbital groove on the orbital surface of the maxilla and then becomes the infra-orbital canal, to emerge on the anterior aspect of the maxilla as the infra-orbital foramen. This groove, canal and foramen successively transmit the infra-orbital nerve (V''), artery and vein (Fig. 197).

The medial wall of the orbit is made up principally by the thin-walled orbital plate of the ethmoid and the lacrimal bone. The ethmoid here is translucent and the ethmoid air cells can be seen through it in the dried specimen. On its superior border with the frontal bone can be seen the anterior and posterior ethmoid foramina.

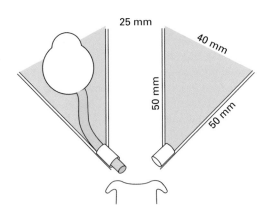

Fig. 224 Schema of orbits.

The orbital surface of the lacrimal bone bears the lacrimal groove, demarcated by the prominent posterior, and less well marked anterior, lacrimal crest. The posterior crest marks the attachment of the medial check ligament of the fascial sheath of the eye, Tenon's fascia (see below). Operations on the lacrimal sac are thus conducted on the superficial aspect of this fascial layer, which protects the orbital contents.

The anterior edge of the lacrimal bone meets with the posterior border of the frontal process of the maxilla to complete the bony fossa for the lacrimal sac.

The smallest contribution to the bony framework of the orbit is the orbital process of the palatine bone, which lies at the medial junction of the superior and inferior orbital fissures, wedged between the ethmoid and the orbital surface of the maxilla.

The periosteum lining the orbit is rather loosely attached to the underlying bone. Parts of the bony walls of the orbit can be surgically removed, therefore, without opening into the orbital cavity proper.

The orbital foramina

These can be listed as follows:
1   Optic foramen; transmitting the optic nerve (II) and ophthalmic artery.
2   Superior orbital fissure; III, IV, V', VI, superior and inferior ophthalmic veins.
3   Inferior orbital fissure; infra-orbital nerve and vessels.
4   Nasolacrimal canal; lacrimal sac.
5/6 Posterior and anterior ethmoidal; transmit ethmoidal branches of the nasociliary nerve and vessels to the cranial cavity and nasal cavity, respectively.
7   Supra-orbital notch; supra-orbital nerve (V') and vessels.
8/9 Zygomaticofacial/zygomaticotemporal; minute canals for nerves of the same name (V'').

The subdivisions of the orbit

The orbit is divided into four areas (Fig. 225):
1   The *eyeball*.
2   The *preseptal space*, defined as anterior to a vertically aligned orbital septum. This is a thin sheet of connective tissue that encircles the orbit as an extension of the periosteum (see below). The septum provides an important barrier to the anterior or posterior extravasation of blood or spread of infection.
3   The *retrobulbar (intraconal) space*, defined as the area posterior to the septum and within a ring formed by the rectus muscles. The retrobulbar space encloses cranial nerves II, III and VI, the nasociliary nerve, the autonomic ciliary ganglion and the ophthalmic vessels. As a consequence, the injection of local anaesthetic into this space (retrobulbar injection) provides rapid and effective anaesthesia.

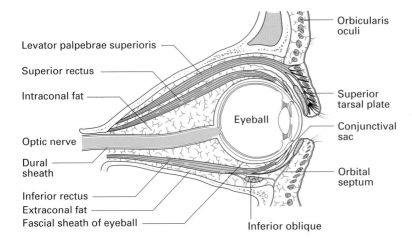

Levator palpebrae superioris

Superior rectus

Intraconal fat

Eyeball

Optic nerve

Dural sheath

Inferior rectus

Extraconal fat

Fascial sheath of eyeball

Inferior oblique

Orbicularis oculi

Superior tarsal plate

Conjunctival sac

Orbital septum

Fig. 225
Compartments of the orbit.

4 The *peribulbar space* is defined as an area posterior to the septum and lying outside the rectus cone. This space contains fewer nerves: the lacrimal and frontal (V′), and trochlear (IV). As a consequence, injection of local anaesthetic into this area requires a larger volume and acts by diffusion into the retrobulbar space.

The eyeball (Fig. 226)

The eyeball, which is just under 25 mm in all diameters, is formed by segments of two spheres of different sizes: a prominent anterior segment, which is transparent and forms about one-sixth of the eyeball, and a larger posterior segment, which is opaque and comprises five-sixths of a sphere. The optic nerve enters the eye approximately 3 mm to the nasal (medial) side of the posterior pole.

The eyeball is formed by three coats: a fibrous outer coat, a vascular middle coat and an inner neural coat – the retina.

Light passes through the cornea, anterior chamber, pupil, posterior chamber, lens and vitreous chamber before striking the retina. Interestingly, light passes through the entire retina before striking the retinal photoreceptors. The photoreceptors then initiate a chain of impulses to bipolar and then ganglion cells. The ganglion cells form the optic nerve, which traverses the orbit to the brain (Figs 183, 184).

The fibrous coat

The fibrous coat comprises a transparent anterior part (the cornea) and an opaque posterior portion (the sclera). Peripherally, the cornea is continuous with the sclera at the sclerocorneal junction. The sclera is a tough, fibrous membrane that is responsible for the maintenance of the shape of the eyeball and that receives the insertion of the extra-ocular muscles. Posteriorly, it is pierced by the optic nerve, with whose dural sheath it is continuous.

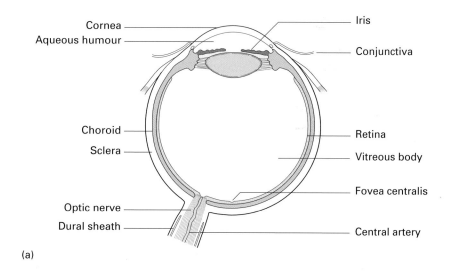

(a)

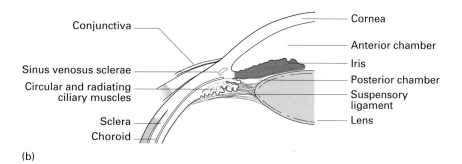

(b)

Fig. 226 (a) The eyeball in section; (b) detail of the ciliary region.

The vascular coat

This is made up of the choroid, the ciliary body and the iris.

The *choroid* is a thin but highly vascular membrane lining the inner surface of the sclera. Posteriorly it is pierced by the optic nerve, and anteriorly it is connected to the iris by the ciliary body.

The *ciliary body* includes the ciliary ring (a fibrous ring continuous with the choroid), the ciliary processes (a group of 60–80 folds arranged radially between the ciliary ring and the iris and connected posteriorly to the suspensory ligament of the lens) and the ciliary muscles (an outer radial and inner circular layer of smooth muscle responsible for the changes in convexity of the lens in accommodation and supplied by parasympathetic fibres transmitted in the oculomotor nerve, III).

The *iris* is the contractile disc surrounding the pupil. It consists of four layers:

1 an anterior mesothelial lining;

2 a connective tissue stroma containing pigment cells;

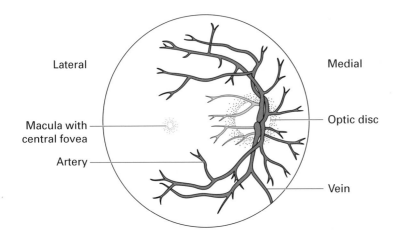

Lateral

Medial

Macula with
central fovea

Optic disc

Artery

Fig. 227 The right
fundus oculi as seen
through the
ophthalmoscope.

Vein

3 a group of radially arranged smooth muscle fibres – the dilator of the
pupil (supplied by the sympathetic system) – and a circular group – the
papillary sphincter (supplied by the parasympathetic fibres in the oculomotor nerve);
4 a posterior layer of pigmented cells that is continuous with the ciliary
part of the retina.

The neural coat

The retina is formed by an outer pigmented and an inner nervous layer,
and is interposed between the choroid and the hyaloid membrane of the
vitreous. Anteriorly, it presents an irregular edge, the *ora serrata*, while posteriorly the nerve fibres on its surface collect to form the optic nerve. Its
appearance as seen through an ophthalmoscope is shown in Fig. 227. Near
its posterior pole there is a pale yellowish area, the *macula lutea*, the site
of central vision, and just medial to this is the pale *optic disc* formed by
the passage of nerve fibres through the retina, corresponding to the 'blind
spot'. The *central artery of the retina* emerges from the disc and then divides
into upper and lower branches; each of these in turn divides into a nasal
and temporal branch. Histologically, the retina consists of a number of layers, but, from a functional point of view, only three need be considered: an
inner receptor cell layer – the layer of rods and cones; an intermediate layer
of bipolar neurones; and the layer of ganglion cells, whose axons form the
superficial layer of optic nerve fibres (Fig. 182).

Contents of the eyeball

Within the eyeball are found: the lens, the aqueous humour and the
vitreous body.
   The *lens* is biconvex and is placed between the vitreous and the aqueous
humour, just behind the iris. Opacities of the lens (cataracts) may obstruct
vision, especially if placed in the central or posterior aspect of the lens.

The *aqueous humour* is a filtrate of plasma secreted by the vessels of the iris and ciliary body into the *posterior chamber* of the eye, i.e. the space between the lens and the iris. From here, it passes through the pupillary aperture into the *anterior chamber* (between the cornea and the iris) and is reabsorbed into the ciliary veins by way of the *sinus venosus sclerae* (or *canal of Schlemm*).

The vitreous body, which occupies the posterior four-fifths of the eyeball, is a thin transparent gel contained within a delicate membrane – the *hyaloid membrane* – and pierced by the lymph-filled *hyaloid canal*. The anterior part of the hyaloid membrane is thickened, receives attachments from the ciliary processes and gives rise to the *suspensory ligament of the lens*. This ligament is attached to the capsule of the lens in front of its equator and serves to retain it in position. It is relaxed by contraction of the radial fibres of the ciliary muscle and so allows the lens to assume a more convex form in accommodation (close reading).

The orbital muscles (Figs 183, 225, 228)

These are the levator palpebrae superioris and the extra-ocular muscles: the medial, lateral, superior and inferior recti and the superior and inferior obliques. The four *recti* arise from a tendinous ring around the optic

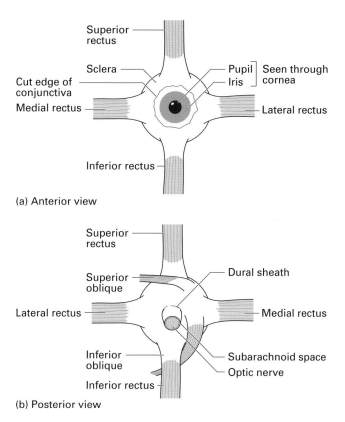

Fig. 228 Orbital muscles of the right eye in schematic views. (a) Anterior aspect; (b) posterior aspect.

(a) Anterior view

(b) Posterior view

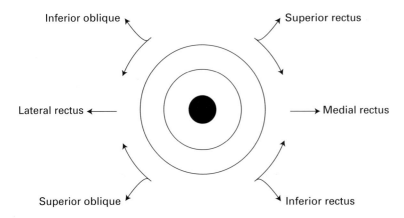

Inferior oblique   Superior rectus

Lateral rectus   Medial rectus

Superior oblique   Inferior rectus

Fig. 229 The direction of action of the muscles acting on the right eyeball from the primary position, i.e. looking directly forward.

foramen and the medial part of the superior orbital fissure and are inserted into the sclera anterior to the equator of the eyeball. The lateral rectus is supplied by the abducent (VI) nerve; the others by the oculomotor (III) nerve. The *superior oblique* arises just above the tendinous ring and is inserted by means of a long tendon that loops around a fibrous pulley on the medial part of the roof of the orbit into the sclera just lateral to the insertion of the superior rectus. It is supplied by the trochlear (IV) nerve. The *inferior oblique* passes like a sling from its origin on the medial side of the orbit around the undersurface of the eye to insert into the sclera between the superior and lateral recti; it is supplied by III. Both the oblique muscles insert behind the equator of the eyeball.

The eyeball is capable of elevation, depression, adduction, abduction and rotation. The medial and lateral recti move the eyeball in one axis only. The four other muscles move it in all three axes:

1 superior rectus: elevation, adduction and medial rotation (or intorsion);
2 inferior rectus: depression, adduction and lateral rotation (or extorsion);
3 superior oblique: depression, abduction and medial rotation;
4 inferior oblique: elevation, abduction and lateral rotation.

Pure elevation and depression of the eyeball is produced by one rectus acting with its opposite oblique – superior rectus with inferior oblique producing pure elevation and inferior rectus with superior oblique producing pure depression. A useful mnemonic is that the superior oblique is 'the tramp's muscle' – it moves the eye 'down and out'. The actions of these muscles are shown in Fig. 229.

The fascial sheath of the eye

A thin fascial membrane, the *vagina bulbi*, or *Tenon's fascia*, ensheaths the eyeball from the corneo-scleral junction to the optic nerve; here, it fuses with the dural sheath of the nerve as it enters the eyeball. This fascia separates the eyeball from the surrounding orbital fat, which lies between it and the ocular muscles. The tendons of these muscles perforate the fascial membrane, which is reflected onto each of these muscles as its fascial

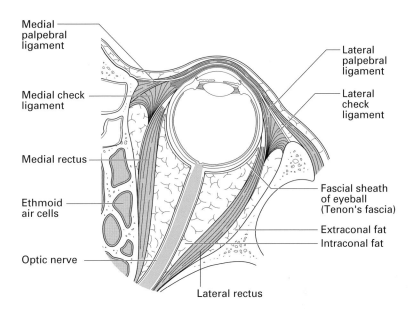

Fig. 230 The orbit in transverse section.

sheath. The fascial sheaths of the recti are thickened anteriorly and, just before they blend with the vagina bulbi, they form a distinct fascial ring. This is thickened inferiorly to form the *suspensory ligament of the eye*. This has extensions medially and laterally that attach to the orbital bony walls as the *medial* and *lateral cheek ligaments*, thus producing a hammock-like sling to support the eyeball (Fig. 230).

The eyelids and conjunctiva (Fig. 231)

Of the two eyelids, the upper is the larger and more mobile, but, apart from the presence of the levator palpebrae superioris in this lid, the structure of the eyelids is essentially the same. Each consists of the following

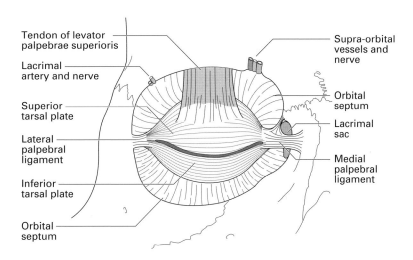

Fig. 231 Dissection of the eyelid.

layers, from without inwards: skin, loose connective tissue, fibres of the orbicularis oculi muscle, the tarsal plates of very dense fibrous tissue, tarsal glands and conjunctiva. The eyelashes arise along the mucocutaneous junction and immediately behind the lashes there are the openings of the *tarsal (Meibomian) glands*. These are large sebaceous glands whose secretion helps to seal the palpebral fissure when the eyelids are closed and forms a thin layer over the exposed surface of the open eye. If blocked, they distend into Meibomian cysts.

The medial ends of the tarsal plates attach by the strong *medial palpebral ligament* to the anterior crest of the lacrimal bone and the adjacent frontal process of the maxilla in front of the lacrimal sac. The *lateral palpebral ligament* passes to the zygomatic bone just within the orbital margin (Fig. 231). The orbital septum is a thin fibrous fascial sheath attached to the periosteum of the orbital rim. In the upper lid, it blends with the fascia over levator palpebrae superioris; in the lower lid, it blends with the margins of the inferior tarsal plate. In front of the orbital septum lies the preseptal space.

The sensory innervation of the upper lid originates from the supra-orbital, supratrochlear and lacrimal nerves – all three are branches of the ophthalmic nerve (V'). The sensory innervation of the lower lid originates from the infra-orbital branch of the maxillary nerve (V'') and infratrochlear (V') nerves. These nerves lie outside the rectus cone and are therefore ineffectively blocked by retrobulbar injections. Preseptal anaesthesia is more effective in providing anaesthesia to the eyelids and conjunctiva.

The *conjunctiva* is the delicate mucous membrane lining the inner surface of the lids, from which it is reflected over the anterior part of the sclera to the cornea, where it becomes continuous with the corneal epithelium. Over the lids it is thick and highly vascular, but over the sclera it is much thinner. The line of reflection from the lid to the sclera is known as the conjunctival fornix; the superior fornix receives the openings of the lacrimal glands.

Movements of the eyelids (superior much more than inferior) are brought about by the contraction of the orbicularis oculi and levator palpebrae superioris muscles. The width of the palpebral fissure at any one time depends upon the tone of these muscles and the degree of protrusion of the eyeball.

The lacrimal apparatus (Figs 231, 232)

The *lacrimal gland* is situated in the upper, lateral part of the orbit in what is known as the lacrimal fossa. The main part of the gland is about the size and shape of an almond, but is connected to a small terminal process, the palpebral lobe, which extends into the posterior part of the upper lid. The gland is drained by a series of 8–12 small ducts that pass through the palpebral lobe into the lateral part of the superior conjunctival fornix, whence its secretion is spread over the surface of the eye by the action of the lids.

The tears are drained by way of the *lacrimal canaliculi*, whose openings, the *lacrimal puncta*, can be seen on the small elevation near the medial margin of each eyelid known as the *lacrimal papilla*. The two canaliculi,

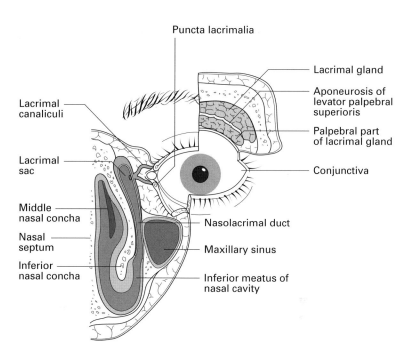

Fig. 232 The lacrimal gland and its drainage system.

superior and inferior, open into the *lacrimal sac*, which is situated in a small depression on the medial surface of the orbit. This in turn drains through the nasolacrimal duct into the anterior part of the inferior meatus of the nose. The nasolacrimal duct, which not uncommonly becomes obstructed, is approximately 12 mm in length and lies in its own bony canal in the medial wall of the orbit.

CLINICAL NOTE

Local anaesthesia for eye surgery

In the past the goal of local anaesthesia for surgery on the eye was analgesia of the globe, akinesia of the external ocular muscles and reduced intraocular pressure. Widespread use of phacoemulsification (emulsification of the lens with an ultrasonic handpick and aspiration) has reduced the need for retrobulbar blocks. Most ophthalmic surgery is performed on the anterior segment and may be performed without retrobulbar block, which has a higher rate of complications. A detailed anatomical knowledge of the target area will allow for appropriate selection of block for specific clinical situations. The three techniques in current usage are: retrobulbar block, peribulbar block and sub-Tenon's block.

Sub-Tenon's block is the most commonly used. In this technique, the conjunctiva and Tenon's capsule are grasped with forceps and a small incision is made. A blunt, curved needle or cannula is then inserted until its tip is at the equator of the globe or even more posterior to this. Small volumes of local anaesthetic (2–5 ml) injected at this site will provide analgesia of the globe by spreading around it under Tenon's fascia. The injection of larger volumes will encourage spread of the local anaesthetic into the sheaths of the extra-ocular muscles as they penetrate the fascia, promoting blockade of the motor nerves within the sheaths.

A retrobulbar block aims to place local anaesthetic solution within the cone, and is therefore perhaps more correctly called intraconal anaesthesia. With insertion of a needle and local anaesthetic in this space comes an increased risk of complications that include intravascular injection, retrobulbar haemorrhage, direct trauma to the optic nerve and spread of local anaesthetic along the optic nerve sheath or dural sheath to the brainstem. For this reason, many anaesthetists now favour peribulbar, or extraconal, injection of local anaesthetic. This technique is not without complications, even in experienced hands, but the incidence is thought to be lower than with intraconal injections.

The abdominal wall

The detailed anatomy of the abdominal wall is essential knowledge to any anaesthetist who proposes to perform safe and effective regional blocks in this area.

Landmarks

A few fixed points enable surface markings to be correlated with deeper structures (Fig. 233).

The *xiphoid* lies opposite the body of the 9th thoracic vertebra.

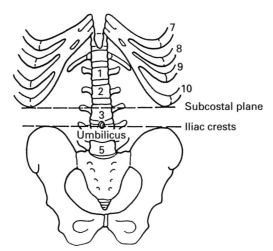

Fig. 233 Boundaries, bony landmarks and vertebral levels of the abdomen.

The costal margin extends downwards and outwards from the xiphoid to the lower margin of the 10th rib, beyond which the 11th and 12th ribs are not long enough to extend. This lower border of the thoracic cage (the *subcostal plane*) is at the level of the 3rd lumbar vertebra.

The *transpyloric plane*, which passes through the 1st lumbar vertebra, lies halfway between the suprasternal notch and the pubis; it is good enough to regard this level as a hand's breadth below the xiphoid.

A line joining the top of the iliac crests usually passes through the 4th lumbar vertebra; it is thus a convenient safety level for lumbar puncture, since the spinal cord usually terminates at the L1–2 junction in the adult.

The umbilicus lies at the level of the L3–4 interspace. It is an inconstant landmark and is lower in the child, in the obese and in the pregnant.

Fascia

There is no deep fascia over the trunk – its presence would render deep breathing or abdominal distension impossible. The superficial fascia, or fat, over the lower abdomen is, however, more fibrous on its deep aspect. It is customary to term the more superficial fatty subcutaneous tissue *Camper's fascia* and the deeper fibrous tissue *Scarpa's fascia*. There is no true anatomical differentiation into these two separate layers.

Muscles (Figs 47, 234, 235)

Rectus abdominis has a 2.5 cm wide origin from the pubic crest and symphysis and a 7.5 cm wide insertion into the 5th, 6th and 7th costal cartilages. The segmental developmental origin of this muscle is commemorated by three fibrous intersections that are situated at the level of the umbilicus, the level of the xiphoid and halfway between the two; a fourth intersection is sometimes found below the umbilicus. These intersections are present only on the anterior aspect of the muscle, where each adheres to the anterior rectus sheath. Injected local anaesthetic is thus prevented from free dissemination throughout the anterior compartment of the rectus sheath, although it will spread without hindrance in the posterior sheath.

Pyramidalis is a small, inconstant, triangular muscle that arises from the pubis, lies in front of the rectus and is inserted into the linea alba.

The *rectus sheath*, within which lies the rectus, is formed, in the main, by a split in the internal oblique aponeurosis. Posteriorly, this sheath is reinforced by the aponeurosis of transversus abdominis, and anteriorly by that of the external oblique.

This basic arrangement is altered at both extremities (Fig. 236). Above the costal margin, the rectus lies directly on the costal cartilages; the anterior sheath here consists only of the external oblique aponeurosis, simply because neither internal oblique nor transversus extends above the rib margin. For 5–8 cm below the costal margin, transversus abdominis remains muscular almost to the midline; these muscle fibres can be seen

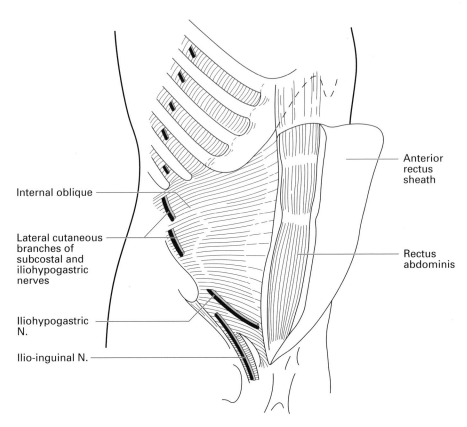

Fig. 234 The anterior rectus sheath has been opened and reflected medially, the external oblique removed and the chest wall dissected down to the internal intercostal muscles.

distinctly in the posterior wall of the upper part of the rectus sheath. Below a level halfway between the umbilicus and pubis, demarcated by the arcuate line of Douglas, the aponeuroses of all three lateral muscles pass in front of rectus. Here, then, posteriorly the rectus rests against transversalis fascia, extraperitoneal fat and peritoneum.

The posterior rectus sheath can thus be said to be, from above downwards:

1 cartilaginous (above the costal margin);
2 muscular (where the muscle fibres of transversus obtrude into the posterior sheath);
3 aponeurotic (the main bulk of the sheath);
4 areolar (below the arcuate line of Douglas).

The resistance of the tough fascia of the anterior wall of the rectus sheath is easily appreciated throughout its extent by the anaesthetist's needle. The posterior wall is less readily defined both above the pubis, where the sheath is deficient, and also high in the epigastrium, where it is softly muscular and not a tough, fibrous aponeurosis.

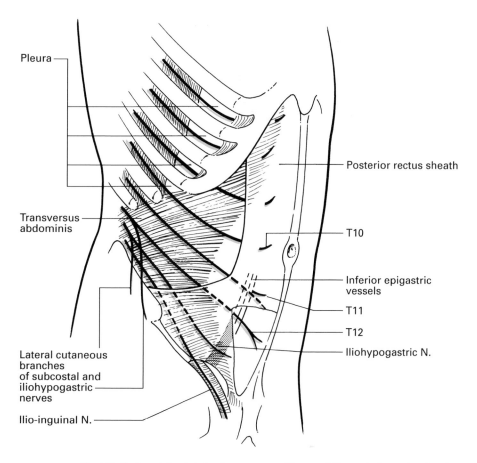

Pleura

Posterior rectus sheath

Transversus abdominis

T10

Inferior epigastric vessels

T11

T12

Iliohypogastric N.

Lateral cutaneous branches of subcostal and iliohypogastric nerves

Ilio-inguinal N.

Fig. 235 Rectus abdominis has been removed to reveal the posterior rectus sheath. In the upper abdomen a window has been cut out of the internal oblique to demonstrate the lower intercostal, subcostal and 1st lumbar nerves lying on the transversus abdominis. The innermost intercostal muscle layer is shown in each intercostal space.

The aponeuroses that form the rectus sheath fuse from pubis to xiphoid in the almost avascular midline *linea alba*. This is narrow and quite difficult to define in the lower abdomen but broadens out considerably above the umbilicus.

The *three lateral muscles of the abdominal wall* fill the space between the rectus in front, the lumbar muscles behind, the costal margin above and the iliac crest below. Their medial extensions constitute the rectus sheath, as described above, and then fuse into the linea alba in the midline.

Above the level of the iliac crest, the fibres of *external oblique* pass downwards and medially, those of *internal oblique* pass upwards and medially and those of *transversus abdominis* pass transversely. Below this level, all the muscles are aponeurotic and all their fibres pass downwards and medially in the formation of the inguinal canal.

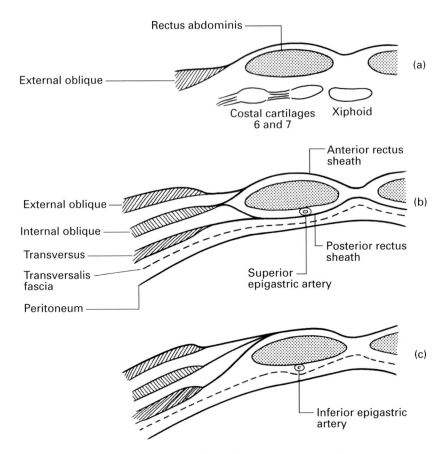

Fig. 236 The composition of the rectus sheath shown in transverse sections: (a) above the costal margin; (b) above the arcuate line; (c) below the arcuate line.

The muscles of the abdominal wall are accessory muscles of respiration that come into play in forcible expiration and coughing; they act by raising the intra-abdominal pressure and also by drawing the lower ribs downwards and medially. Electromyographic studies have shown that the abdominal muscles are not used in inspiration.

CLINICAL NOTE

Rectus sheath blocks

Analgesia can be provided with rectus sheath blocks for operations that include incisions in the anterior abdominal wall. Although a number of technique variations exist, the principle is that local anaesthetic is placed within the rectus sheath, aiming to deposit it slightly anterior to the posterior sheath. Multiple injections are used, both above and below the level of the umbilicus and between 2 and 5 cm lateral to it. The penetration of the needle through the anterior rectus sheath can be appreciated as a definite 'pop' if a short-bevel or blunted needle is used. Identification of the posterior rectus sheath can be achieved if the needle is moved back and forth and a characteristic 'scratching' sensation is felt. Care must be exercised with the doses of local anaesthetic used, as in any multiple-injection technique, particularly if rectus sheath blocks are combined with ilio-inguinal blocks for low abdominal incisions.

Blood supply

There is a rich blood supply to the abdominal wall; its details are unimportant to the anaesthetist, except for the position of the inferior and superior epigastric vessels, which lie in the posterior rectus sheath and which may be wounded in performing a rectus block. The surface marking of these vessels is a line that curves gently from the femoral pulse at the groin to a point one finger's breadth lateral to the umbilicus and that runs thence vertically upwards to the costal margin. However, considerable variations may occur and, moreover, there are frequently large branches of the artery that pass superomedially within the posterior sheath. Any of these may be torn by a needle or tracer.

The *inferior epigastric artery* is derived from the external iliac artery, skirts medially to the internal inguinal ring and enters the posterior rectus sheath beneath the arcuate line of Douglas.

The *superior epigastric artery* is smaller; it enters the upper part of the rectus sheath behind the 7th costal cartilage as a terminal branch of the internal thoracic artery, runs vertically downwards and anastomoses with the inferior artery.

Nerve supply (Figs 46, 47, 234, 235)

The abdominal wall is innervated by the anterior primary rami of T7–L1. Its segmental cutaneous supply is readily mapped out if it is remembered that T7 supplies the xiphoid, T10 the umbilicus and L1 the groin.

The *intercostal nerves*, T7–11, and the *subcostal nerve*, T12, enter the abdominal wall between the interdigitations of the diaphragm and transversus abdominis (Fig. 235). The intercostal nerves maintain the same relationship to the muscles of the abdominal wall as they have with the intercostal muscles. In their thoracic course, they lie between the second and third layers of intercostal muscles (the internal intercostals and innermost intercostals); in their progress between the lateral abdominal muscles, they lie between the second and third layer, the internal oblique and transversus abdominis, as shown in Fig. 235. In this plane, the nerves are conducted medially behind the rectus, which they then pierce to supply the overlying skin.

In contrast, the 1st lumbar nerve divides in front of quadratus lumborum into the iliohypogastric and ilio-inguinal nerves, which penetrate the transversus abdominis to lie between transversus and internal oblique.

The *iliohypogastric nerve* pierces the internal oblique immediately above and in front of the anterior superior iliac spine, runs deep to the external oblique, just superior to the inguinal canal, and ends by supplying the suprapubic skin.

The *ilio-inguinal nerve* also pierces the internal oblique and then traverses the inguinal canal in front of the spermatic cord. It emerges either through the external ring itself or through the adjacent external oblique aponeurosis to supply the skin of the scrotum (or labium majus) together with the adjacent upper thigh.

Each nerve apart from the ilio-inguinal nerve gives off a lateral cutaneous branch in the mid-axillary line. These branches from the intercostal nerves T7–11 divide into an anterior and posterior branch supplying the skin from the lateral edge of rectus in front to the erector spinae behind. The lateral cutaneous branches of both the subcostal (T12) and iliohypogastric nerve do not divide but run downwards to supply the skin over the upper lateral aspect of the buttock (Fig. 47).

Each nerve T7–12 gives off a small collateral branch that runs parallel with it; by analogy, the ilio-inguinal nerve is to be regarded as the collateral branch of the iliohypogastric nerve, and hence has no lateral cutaneous branch.

CLINICAL NOTE

Ilio-inguinal, iliohypogastric and genitofemoral nerve blocks

Blockade of the ilio-inguinal and iliohypogastric nerves can provide anaesthesia for, or analgesia after, inguinal hernia repair. If anaesthesia is required, a block of the genital branch of the genitofemoral nerve will also be needed. The ilio-inguinal and iliohypogastric nerves can be blocked as follows: a short-bevel or blunted needle is passed through the skin at a point 2 cm medial and 2 cm inferior to the anterosuperior iliac spine, aiming towards the pubis at an angle of 45–60°. The passage of the needle tip through the external oblique aponeurosis can be appreciated as a 'pop' or sudden loss of resistance to the passage of the needle. Local anaesthetic (7–10 ml) is then injected. The needle is then passed a further 1–2 cm through the softer resistance of the internal oblique muscle. A further 7–10 ml of local anaesthetic is injected. The genital branch of the genitofemoral nerve can be blocked by depositing 10 ml of local anaesthetic just lateral to the pubic tubercle below the level of the inguinal ligament.

Part 8
The Anatomy of Pain

Introduction

Much of the knowledge of the anatomy and neurophysiology of pain comes from animal studies and clinical observations in humans with damaged nervous systems. This chapter is not, like most anatomical texts, based on the observations of the anatomist but more on the work of neurophysiologists and clinicians and, most recently, on the work of scientists developing new methods of imaging of the central nervous system. The search for the anatomy of pain has a long history. In 1644, Descartes postulated the existence of delicate threads that linked the periphery to the brain and that facilitated the perception and localization of pain, initiating the search for pain centres and pathways that has continued for 300 years. The conception of pain as a mere aspect of sensation has been challenged by recent advances in imaging such as positron emission tomography (PET) and functional magnetic resonance imaging (fMRI). PET scanning has demonstrated that changes in local cerebral blood flow (CBF) reflect variations in synaptic activity, and fMRI signal changes in proportion to CBF.

The meaning of pain in an evolutionary sense is to inform the organism of tissue damage. There is no anatomically discrete Descartian pain 'pathway' – the system is not hard wired, so to speak. The 'pathway' exhibits plasticity, i.e. functional changes occur in the transmission system as a result of its activation by tissue damage or inflammation. In response to tissue damage, there is a dynamic interlocking series of biological interactive mechanisms that result in vasodilatation, swelling, activation of inflammatory mediators and, via the central nervous system, the perception of pain.

Mediators released by cell damage and produced by noxious stimuli produce electrophysiological activity in nociceptive primary afferent nerve fibres. After this immediate response, there is transport of chemicals such as neurotrophins along axons to the cell bodies of the dorsal root ganglia, where these chemicals alter the chemistry and physiology of the sensory cell. Sensory impulses are modulated both at the spinal level by local cellular circuitry of the dorsal horn of the spinal cord and also by descending control systems originating in the cortex, thalamus and brainstem. Sensory impulses are transmitted to various parts of the central nervous system, where registration and localization occurs, and affective responses are produced. There is no discrete pain centre within the brain; responses to pain are produced through integrated activity of various areas of the brain, *the pain matrix*.

Pain has been defined as 'an unpleasant sensory and emotional experience associated with actual or potential tissue damage or described in terms of such damage'. It is thus a *subjective* experience, and subject to significant modulation by higher centres, particularly with regard to the emotional or affective component. Human beings attribute various adjectives to pain that are often a guide to its aetiology. Pain may be 'experienced' as a result of tissue trauma, such as occurs after surgery; through damage to the processing system itself, as can occur after spinal cord injury and in peripheral neuropathy. In some chronic pain states there may be no demonstrable pathology (identifiable with current levels of technology) in either the nervous system or the tissues. Activation of peripheral receptors that respond to noxious stimuli (mechanical, chemical or thermal energy) initiates the conscious perception of pain subject to central processing and modulation. The activation of these receptors, called nociceptors, is termed *nociception*. However, pain can be experienced without activation of these receptors; thus, pain can exist *without* tissue damage.

Classification of pain

Pain can be broadly divided on a functional basis into 'nociceptive pain', in which the experience occurs as a result of activation of peripheral nociceptors, and 'neuropathic pain', in which there is dysfunction or injury to the nervous system itself with or without tissue damage. This distinction has some implications for therapy. Nociceptive pain is usually regarded as being more opioid-sensitive than neuropathic pain, which, generally speaking, is thought to be more responsive to non-opioid analgesics such as anticonvulsants and antidepressants. This division on the basis of opioid response is now not regarded as being so clearly defined.

Pain may also be classified on the basis of where it is felt. Superficial pain is produced by tissue damage to the skin and to the mucosa of the mouth and anus. It is very well localized and sharp. Deep pain is felt both from somatic structures, where it is often described as dull and aching but can be localized within the segmental nerve supply, and from visceral structures, where localization is poor.

Peripheral receptors and afferent fibres

The axons of primary afferent nerves end in skin, subcutaneous tissue, periosteum, joints, muscles and viscera. They ramify profusely and near their termination the perineural sheaths are shed. These peripheral free nerve endings are termed nociceptors, and will respond to chemical, mechanical or thermal energy. The axons of primary afferent neurones involved in nociception are classified into two major groups on the basis of axon size, myelination and conduction velocity. They are titled Aδ and C. A summary of characteristics is given in Table 4.

About 25% of Aδ fibres and a variable proportion of the C fibres specifically respond to the activation of nociceptors. The rest of the fibres respond to low-intensity stimuli. Aδ fibres are responsible for the fast pain experienced in the first 50 ms after a noxious stimulus. Functionally, the Aδ free endings detect potentially hazardous mechanical and thermal stimuli,

Table 4 Summary of characteristics of the axons of primary afferent neurones involved in pain

Fibre type	Diameter (μm)	Speed (m/s)	Nociceptor?	Respond to	Sensitized by ?	
C	<1.5	0.5–2.0	Variable	Mechanical energy Thermal energy	Leukotrienes, prostaglandin, substance P	Prolonged Burning pain
Aδ	1–4μ	5–30	25%	Mechanical energy Thermal energy	Cyclic adenosine monophosphate, protein kinase A, bradykinin, prostaglandins, neural growth factor, leukotrienes	Brief Sharp pain

and trigger rapid nociception and protective reflex responses. Thus, C fibres, while also reinforcing the immediate response of the Aδ fibres to stress, continue to be stimulated by damaged and inflamed tissues. The primary nociceptive afferents supply skin, subcutaneous tissue, musculature, periosteum, joints and viscera. Some afferents respond to noxious mechanical, chemical and thermal stimuli and are termed polymodal nociceptors. There are also mechano-nociceptors and mechano-thermal nociceptors.

Peripheral sensitization

When tissue damage and inflammation occur after the acute phase of 'pain', C fibres may become sensitized and activated by the presence of inflammatory mediators such as prostanoids, bradykinin and serotonin, among others. Neural growth factor and cytokines may also affect peripheral receptor sensitivity. These chemicals reduce the threshold for stimulation by affecting transduction proteins, which are the mechanism for converting a stimulus into electrical energy. This provides the mechanisms both for tenderness after injury and for development of peripheral sensitization, which leads to a persisting afferent input into the cord that, in turn, may lead to central hyperexcitability. In the early phase, there occur changes to the existing proteins; later, there is alteration of the proteins produced, which may, for example, dramatically alter the activity in ion channels.

Cutaneous nociceptors

In skin, subcutaneous tissue and fascia, all three types of nociceptor described above have been demonstrated, along with Aδ heat, and Aδ and C cold nociceptors. C-polymodal nociceptors account for 95% of C fibres in skin.

Other somatic structures

Muscle is supplied by both Aδ and C fibres, whose free nerve endings within the muscle itself respond to stimuli. The greater part of muscle pain is mediated by unmyelinated C fibre endings, which have a high threshold to mechanical stress but in particular will respond to any ischaemic muscle injury. Periosteum is innervated by both Aδ and C fibres, and the threshold for noxious stimulation is low compared with other structures. Joints are supplied with Aδ and C fibre afferents, some of which end in specialized receptors for torque. However, the majority form a plexus of free nerve endings in the joint cavity.

Visceral nociception

Specific nociceptors in viscera have been identified in the urinary bladder. Visceral pain is characterized by poor localization, vagueness and a deep aching nature. It is often referred, and this may be due to convergence of visceral and cutaneous afferent fibres onto second-order neurones in laminae V–VIII of the dorsal horn of the spinal cord. Here, the brain is unable to distinguish between the two inputs, and projects the sensation to the somatic structure, or the visceral afferent activity facilitates the passage of somatic sensations that are perceived as pain. It has also been proposed that visceral referred pain is produced at the thalamic level. Visceral nociception is produced by distension, stretching and tearing of hollow viscera, by traction and compression, by inflammation and by any mechanism in which the mediators of nociception mentioned above are released, as in ischaemia. Visceral afferent impulses are conveyed with sympathetic afferent fibres via the cervicothoracic ganglia (thoracic viscera), the coeliac plexus (abdominal viscera) and the inferior hypogastric plexus (pelvic viscera). The afferent fibres from the viscera constitute <10% of fibres in dorsal roots and converge on dorsal horn neurones over a number of segments. The visceral sensory field of each root is much larger than that for a cutaneous nerve; hence, pain from the viscera is poorly localized. Once a visceral pathology affects an area with somatic innervation, more accurate localization is possible; hence, the classical progression of the pain of appendicitis from a vague central abdominal (visceral) pain to right-sided pain over McBurney's point as the parietal peritoneum is inflamed.

The spinal cord and central projections
The dorsal horn

The impulses generated by peripheral nociceptors travel in the peripheral nerves, the cell bodies of which are situated in the dorsal root ganglia. Large and small fibres are randomly situated at first, but as the axons pass through the dorsal root the small (C) fibres become mostly situated on the lateral aspect of the root. In the dorsal root entry zone, those small fibres on the medial side of the root cross and join the medial fibres to enter the dorsolateral (Lissauer's) tract. The larger A fibres travel more medially

Fig. 237 Spinal nerve root and spinal cord: pain transmission. 1, 2, spinothalamic and spinoreticular tracts. 3, Lissauer's tract, dorsal root entry zone. Note fibres ascending (they also descend) before entry. Also note that some fibres (12%) enter via the ventral rather than the dorsal root. 4, dorsal horn, substantia gelatinosa (Rexed lamina 3). 5, Rexed laminae 4, 5, 6.

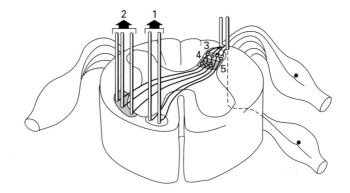

in the dorsal column, although some will enter Lissauer's tract (Fig. 237). Axons may branch both before entering and upon entering the spinal cord, the number of central processes being approximately 40% greater than the number of cell bodies in the ganglia. There is evidence that primary nociceptive afferent axons may also enter the cord via ventral nerve roots and pass dorsally to terminate in the dorsal horn. The proportion of these is unknown but it may explain the failure of dorsal rhizotomy in the treatment of pain.

The spinal grey matter has been divided into 10 laminae on the basis of cyto-architectonic studies. Laminae I–VI make up the dorsal horn, VII–IX make up the ventral horn and lamina X is a cluster of cells around the central canal. Lamina I is termed the marginal layer, lamina II is the substantia gelatinosa (and is divided into IIo (outer) and IIi (inner)) and II–IV is the nucleus proprius (Fig. 238). The laminae run the entire length of the cord,

Fig. 238 The laminae of Rexed with simplified overview of afferent dorsal horn connections. Aδ, mechanoreceptive afferents terminate at laminae III and IV; Aα, nociceptive afferents terminate at laminae I, IIo, V and X; C, afferent fibres terminate at laminae I, IIo and V.

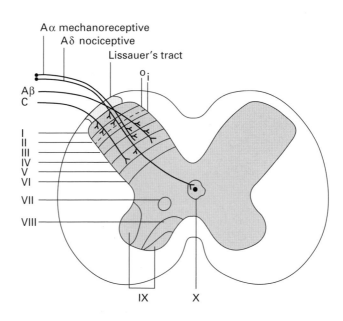

fusing with the medullary dorsal horn. Aδ and C fibres divide on entering Lissauer's tract into an ascending and a descending branch. Aδ fibres
travel one or two segments rostrally or caudally, and C fibre axons spread
three or more segments above and below their point of entry. Aδ fibres
penetrate the lateral aspect of the dorsal horn and terminate at laminae I,
IIo, V and X. C fibres terminate in I, IIo and V. Visceral afferent fibres (sympathetic Aδ and C afferents) and muscle afferents terminate predominately
in laminae I and V. Sympathetic afferents also terminate in the anteromedial lateral cell nucleus. Visceral and cutaneous afferents converge on neurones in laminae I and V, the axons of which mostly project to the contralateral anterolateral fasciculus and thence to the brain. Dorsal horn neurones
project not only to cells that make up ascending systems but also to cells
in the same segment and in other segments of the cord. Descending axons
from inhibitory or facilitatory pathways exert their influence on the dorsal
horn via the dorsolateral funiculus. The cellular circuitry within the dorsal
horn allows for more complex processing of sensory input beyond that of
a simple relay station.

Ascending systems

Nociceptive information is transmitted from the dorsal horn to the thalamus and higher centres mostly via the lateral and ventral spinothalamic tracts. However, nociceptive information is also transmitted via
the spinoreticular tracts, the spinomesencephalic tract (which together
with the spinothalamic tract is called the anterolateral fasciculus) and
the dorsal column postsynaptic spinomedullary system. The spinomesencephalic tract is considered by some to be functionally indistinct from the
spinoreticular tract.

Spinothalamic tract

The spinothalamic tract is the main nociceptive pathway. The cell bodies
of the tract are situated in the dorsal horn in laminae I, V–VIII and IX.
Cell bodies projecting to the ventrobasal and posterior thalamus are concentrated in laminae I and V. Those projecting to the medial thalamus are
found in VI–IX with some in I. The axons cross within one or two segments and pass through the ventral white commisure to ascend as the
lateral spinothalamic tract. The spinothalamic tract neurones from deep
laminae project to the hypothalamus, peri-aqueductal grey matter (PAG),
reticular formation of the medulla, pons and mid-brain, and to medial and
intralaminar thalamic nuclei. Here, these axons synapse with interneu-
rones connected to the limbic and forebrain system. Axons from laminae I
and V project to the ventroposterolateral nucleus and medial posterior thalamus, where they synapse with projections to the somatosensory cortex.
There is a topographic representation both in the spinothalamic tract itself,
with sacral segments laterally and additions coming in medially, and in the
thalamus. Spinothalamic tract fibres are thick, oligosynaptic and rapidly
transmitting. Division of the spinothalamic tract produces hypoalgesia on

the contralateral side, but only in 80–90% of cases. This lack of a complete response may be attributed to uncrossed fibres facilitating transmission of nociceptive impulses.

Spinoreticular tract

The cell bodies of the spinoreticular tract are likely to be in laminae VII and VIII. The axons cross the midline and ascend medial to the spinothalamic tract, synapsing with brainstem reticular formation nuclei. Fibres also project to the smaller intralaminar nuclei of the thalamus. The fibres are thin, relatively slowly conducting and polysynaptic.

The spinothalamic tract and the spinoreticular tract are the principal pathways for nociception in humans. The lateral spinothalamic tract is a more recent development and is much larger in humans than in primates. There is somatotopographic organization and more rapid transmission than in the other ascending systems, and the principal synapse is to the somatosensory cortex. This part of the system seems to be concerned with quantitative analysis and spatial discrimination. The spinoreticular tract and spinomesencephalic tract, on the other hand, project to the older (phylogenetically speaking) parts of the brain and are concerned with reflex responses, endocrine responses and with triggering motivational drive, fear and suffering. Nociceptive input from the sensory cranial nerves travels in the fibres of the bulbothalamic tract, which is distributed with the spinothalamic tract through the nucleus of the trigeminal nerve and descends to the level of the 2nd and 3rd cervical vertebrae.

Cerebral processing and the pain matrix

Projections from the thalamus run to multiple sites during painful stimulation, via a broad network of interacting active areas. This 'matrix' of active areas includes the PAG (an area rich in opioid receptors), the primary and secondary somatosensory cortices, the anterior cingulate cortex, the amygdala and the insula. The pain matrix seems to be divided, in a simplified sense, into medial and lateral systems, representing the part of the thalamus from which the impulse is projected onto higher structures. The lateral system, projecting onto the primary and secondary somatosensory cortex, seems to be concerned with discrimination and localization; the medial system serves to produce the affective component of pain; and the insula seems to occupy a role including both types. One area of the cortex, the frontal operculum, produces a sensation of pain under electrical stimulation. The realization that multiple areas of the cortex (the pain matrix) (Fig. 239) are involved in how humans perceive and react to pain is a critical juncture in the understanding of the biology of pain. Imaging has changed the way we understand pain; fMRI has also demonstrated variations in brain activity in patients with chronic pain compared with normal controls.

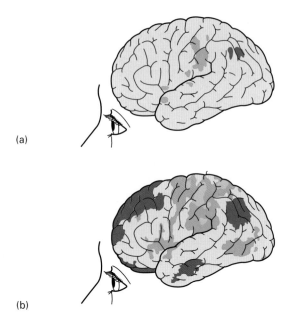

(a)

(b)

Fig. 239 Diagram of the types of changes in functional activity that may be demonstrated by functional magnetic resonance imaging. (a) The subject has no pain; (b) the patient with chronic pain has a different pattern of brain activity.

Modulation of pain signals

The human brain regulates afferent pain input by a variety of mechanisms. This ability of the brain to filter pain signals may be part of the responses to stress – *the flight or fight response*. Absence of pain when tissue damage has occurred has been documented many times on the battlefield. Psychological modification such as using distraction has been shown to modulate the activity of the pain matrix, as has the administration of analgesic drugs such as opioids. Tricyclic antidepressant drugs produce analgesia by their effect on descending modulatory systems. Afferent pain stimuli can thus be affected by exogenous and endogenous mechanisms.

Cortical modulation

Stimulation of the PAG by electrical currents can produce intense analgesia without sensory or motor effects. Intense excitement, distraction or intense motivation, as occur in very high levels of stress, can modulate or abolish the perception of pain in response to nociception. Such experiences are well documented in battlefield and sporting injuries, and the release of endogenous opioid molecules (endorphins) is thought to be responsible. Opioid receptors and endorphins are found in high concentrations in the PAG. Anxiety and depression will also tend to increase the perception of pain in response to a given stimulus. This may be related to decreased discharge in descending inhibitory pathways or altered function in the cortical processing areas. The PAG receives input from the frontal cortex and other cognitive areas, the limbic system, the thalamus and the hypothalamus.

Descending inhibitory pathways

There are three descending inhibitory pathways. First, ascending nociceptive impulses from the anterolateral fasciculus input into the medullary and mesencephalic nuclei (in particular, the nucleus raphe magnus), the medullary pontine and mesencephalic reticular formation, and the PAG via the nucleus cuneiformis. From the PAG, descending inhibitory pathways project via the pons to the dorsal horn of the spinal cord, or via an intermediary synapse in the nucleus raphe magnus, to descend in the dorsolateral funiculus to synapse in the dorsal horn at all the laminae, but particularly I, IIo and V (Fig. 240). Second, there is a direct hypothalamospinal system that descends and sends terminals to the medullary dorsal horn as well as the dorsal horn of the cord, where its terminals are concentrated at laminae I and X. Third, a direct PAG–spinal system bypasses the basal nuclei and projects onto the medullary horn and laminae I, IIo, V and X. These pathways will vary both in their structure and in the neurotransmitter present. The main neurotransmitters involved in descending pain control are thought to be serotonin (5-hydroxytryptamine) and

Fig. 240 Modulation of pain at the spinal cord, medulla and mid-brain. Note that there is initial modulation in the dorsal horn of the spinal cord. Descending modulation is initiated in the peri-aqueductal grey matter, in the medial and lateral reticular formation, and in the nucleus raphe magnus. Descending inhibitory tracts in the dorsolateral fasciculus then impinge on the dorsal horn. (Reproduced with permission from Phillips and Cousins 1986.)

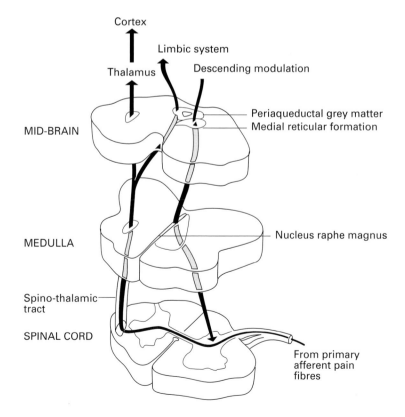

Cortex

Limbic system

Descending modulation

Thalamus

MID-BRAIN

Periaqueductal grey matter
Medial reticular formation

MEDULLA

Nucleus raphe magnus

Spino-thalamic tract

SPINAL CORD

From primary afferent pain fibres

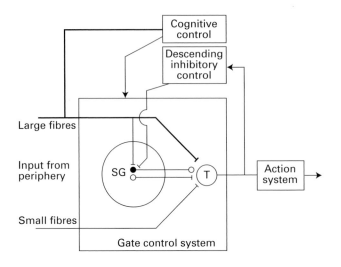

Fig. 241 The gate control theory of pain (mark II). The new model includes excitatory (white circle) and inhibitory (black circle) links from the substantia gelatinosa (SG) to the transmission cells (T) as well as descending inhibitory control from brainstem systems. The round knob at the end of the inhibitory link implies that its action may be presynaptic, postsynaptic or both. All connections are excitatory, except the inhibitory link from the substantia gelatinosa to the transmission cells.

norepinephrine. These descending inhibitory pathways also demonstrate functional plasticity.

The gate control theory of pain

The observation that there was no hard link between stimulus and the degree of pain reported, and that pain perceived was affected by physiological and psychological variables, led Melzack and Wall to propose a gate theory of pain modulation that has subsequently been modified. Large and small fibres input into spinal cord transmission cells and into the substantia gelatinosa (lamina II of Rexed). The transmission of impulses to the transmitter cells is controlled by a spinal gating system. This is in turn influenced by the relative activity in large or small fibres. Activity in large fibres tends to inhibit transmission whereas activity in small fibres tends to stimulate transmission, i.e. opens the gate. The large-diameter fibres are not necessarily nociceptive afferents; flooding the dorsal horn with cutaneous touch and pressure sensation may close the gate for smaller nociceptive input (Fig. 241).

Central sensitization of pain

The effect of primary sensitization is to increase the afferent input into the spinal cord. This along with other mechanisms may produce central sensitization, which is an increase in the activation and lowering of the threshold of excitation of neurones in the central nervous system. This increase in excitability increases the 'gain', such that afferent stimuli that would normally produce a non-painful sensation start to produce pain. There is an early and late phase, the former reflecting changes in synaptic connections in the cord and the latter changes in protein production. Changes occur in descending control systems from the mid-brain and brainstem, and from areas involved with the affective autonomic and aversive aspects of pain. These systems are thought to be serotonergic (serotonin is the neurotransmitter involved). Along with peripheral sensitization there occur reversible physiological and pharmacological changes, which have been described as plasticity and which lead to alterations in the relationship between input stimuli and perceived sensation, fundamental to the development of persisting pain.

The autonomic nervous system and pain

There seem to be many levels in the neuraxis in which the parasympathetic and sympathetic nervous system (SNS) and 'pain' interact, although the SNS seems dominant in this respect (see page 326). In a general sense, the activation of the SNS in the brain suppresses pain and there is a great deal of overlap of cortical regions that respond to painful stimuli and that are activated when the SNS is aroused, such as the anterior cingulate, the insula and the amygdala – areas that have many μ-opioid receptors. SNS-induced analgesia may be seen as part of the stress response through release of endogenous opioids and activation of descending inhibitory pathways.

The mechanism whereby sympathetic blockade produces analgesia is disputed. If the sympathetic nervous system is a purely efferent system, then it should not be involved in afferent impulse transmission. However, there are clear instances, such as in coeliac plexus blockade for foregut malignancies, in which pain relief can be reliably reproduced. Afferent fibres have been demonstrated in sympathetic and parasympathetic nerves supplying the bladder, which would explain analgesia produced by sympathetic nerve blocks. It has also been proposed that there are somatic nerve fibres travelling with the sympathetic nerves that are affected by neural blockade of 'sympathetic trunks' producing analgesia.

Index

Note: This index is in a *classified* form, i.e. arteries, foramen/foramina, ganglia, ligaments, nerves, veins, etc. are all listed under these respective main entries, e.g. 'arteries (named)'. Muscles are an exception, and specific muscles are listed as main entries.

Page numbers in *italics* denote figures, those in **bold** denote tables.